Concrete and Steel Construction Methods

33103-10

Objectives

When you have completed this module, you will be able to do the following:

1. Describe the composition and uses of the common types of commercial building materials.
2. Describe the common methods of commercial construction.
3. Explain common terms used in commercial construction.
4. Identify various types of suspended ceilings.
5. Identify the tools used to make openings in concrete and steel.
6. Select the appropriate drill bits and bore openings in concrete and steel.
7. Select and install appropriate fasteners and anchors in the following:
   - Concrete
   - Steel

Trade Terms

Admixture
American Society for Testing and Materials International (ASTM)
Clearance
Corrugated
Ferrule
Foot-pounds (ft-lbs)
Green concrete
Inch-pounds (in-lbs)
Kerf
Nominal size
Plastic concrete
Plenum
Post-tensioned concrete
Rabbeted
Society of Automotive Engineers (SAE)
Striated
Tolerance
Torque

Prerequisites

Before you begin this module, it is recommended that you successfully complete *Core Curriculum* and *Electronic Systems Technician Level One*, Modules 33101-10 and 33102-10.

EST LEVEL ONE

- **33108-10** Low-Voltage Cabling
- **33107-10** Introduction to the *National Electrical Code®*
- **33106-10** Hand Bending of Conduit
- **33105-10** Craft-Related Mathematics
- **33104-10** Pathways and Spaces
- **33103-10** Concrete and Steel Construction Methods
- **33102-10** Wood and Masonry Construction Methods
- **33101-10** Introduction to the Trade
- **Core Curriculum: Introductory Craft Skills**

This course map shows all of the modules in *Electronic Systems Technician Level One*. The suggested training order begins at the bottom and proceeds up. Skill levels increase as you advance on the course map. The local Training Program Sponsor may adjust the training order.

Contents

Topics to be presented in this module include:

1.0.0 INTRODUCTION

The preceding module covered construction materials and methods, along with the tools and fasteners commonly used in residential construction. The structural framework of larger commercial buildings (*Figure 1*) is made of either concrete or structural steel constructed on a poured concrete foundation. Interior walls are generally framed with cold-rolled steel studs and other steel framing members. Upper floors and roof structures are supported with metal trusses. Floors in many high-rise structures are made of concrete poured into deck forms. Both types of buildings have interior walls and partitions are typically covered with gypsum wallboard.

The fasteners and anchors used for concrete and steel construction are different from those used in wood construction. Some of the tools used to install the fasteners are different as well. In this module, you will learn about the materials, methods, fasteners, and special tools used in concrete and steel construction. This knowledge will assist you when you are installing cabling systems and equipment in commercial structures.

Figure 1 Concrete and steel buildings.

2.0.0 BUILDING MATERIALS

Concrete and steel are the primary building materials in commercial construction. Concrete is used for foundations on all buildings. It is often used to form floors, as well as structural elements. Metal, primarily steel, is used to provide the structural framework of many buildings. Steel is commonly used to build interior framing in commercial structures in the same way that wood is used in residential applications.

2.1.0 Concrete

Concrete is a mixture of four basic materials: portland cement; fine aggregates, such as sand; coarse aggregates, such as stones; and water. When first mixed, concrete is in a semi-liquid state. It is referred to as plastic concrete. When the concrete hardens but has not yet gained structural strength, it is called green concrete. Concrete that has hardened and gained its structural strength is called cured concrete. Different types of concrete can be obtained by varying the basic materials or by adding other materials to the mix. These added materials are called admixtures.

Portland cement is a finely ground powder consisting of varying amounts of lime, silica, alumina, iron, and other trace components. When dry, it may be moved in bulk or bagged in moisture-resistant sacks and stored for relatively long periods of time. Portland cement is a type of hydraulic cement that sets and hardens by reacting with water. It will do so with or without the presence of air. This chemical reaction is called hydration. It can occur even when the concrete is submerged in water. The reaction creates a calcium silicate hydrate gel and releases heat. This reaction begins as soon as water is mixed with the cement. It continues as the mixture hardens and cures. The reaction occurs rapidly at first, depending on how finely the cement is ground and which admixtures are present. After its initial cure and strength have been achieved, the concrete mixture continues to slowly cure over a longer period of time until its ultimate strength is attained.

Concrete is in a semi-liquid form when poured. Therefore, it is placed in reinforced forms made of wood, metal, or other materials (*Figure 2*). Concrete floors, walls, and columns can be poured on-site. Walls and other structural components are sometimes prefabricated off-site and moved by truck to the site where they are lifted into place with cranes.

In residential construction, concrete may be used in foundation walls and footings, basement floors, or as the foundation slab, if the house has no basement.

Figure 2 Concrete being pumped into a wall form.

In commercial construction, the entire structure, including floors, walls, and support columns, may be made of concrete (*Figure 3*). Walls can be anywhere from a few inches to several feet thick.

Building foundations may require thousands of cubic yards of concrete (*Figure 4*).

The ratio of basic ingredients in concrete is determined by a number of variables, such as the application or weather conditions. A common mix for do-it-yourself applications is 3:2:1—three parts aggregate, two parts sand,

Figure 3 Concrete structure.

Figure 4 Concreting a foundation.

Portland Cement Substitute

Concrete is the most common product used in construction. It is used in buildings, roads, bridges, tunnels, and airport runways. Portland cement is a manufactured product that is a key ingredient in concrete. It is also a key ingredient of the mortar used to bind brick and concrete block. The ingredients of portland cement are heated in a kiln, and then ground into a fine dust. This process consumes energy. Scientists have found that fly ash and other industrial by-products can be substituted for up to 60 percent of the portland cement in the concrete mixture. This approach has three positive effects on the environment. First, it reduces the energy consumed in manufacturing portland cement. Second, it reduces consumption of the natural resources needed to manufacture portland cement. Finally, it creates a use for the fly ash and slag that are waste products of coal-burning energy plants and metal foundries.

one part portland cement, and enough water to make the mix workable. In the construction trades, the correct ratio for a given application is usually determined by an engineer. Admixtures may be added to affect drying time, increase strength, and add color.

Those working with cement should be aware that it is harmful. Dry cement dust can enter open wounds and cause blood poisoning. When the cement dust comes in contact with body fluids, it can cause chemical burns to the membranes of the eyes, nose, mouth, throat, or lungs. Wet cement or concrete can also cause chemical burns to the eyes and skin. Make sure that appropriate personal protective equipment is worn when working with dry cement or wet concrete. If wet concrete enters waterproof boots from the top, remove the boots and rinse your legs, feet, boots, and clothing with clear water as soon as possible. Repeated contact with cement or wet concrete can cause an allergic reaction in certain individuals.

2.2.0 Metal

Structural steel provides the supporting framework for many commercial buildings. *Figure 5* shows an example of a building addition framed in structural steel. Steel I-beams are the most commonly used shape. They are shaped like the capital letter I. The decision to use either concrete or structural steel as the building framework is often based on economics.

Cold-rolled steel is used for both interior and exterior framing in commercial buildings. The framing members used in loadbearing walls (*Figure 6*) are of a much heavier gauge than those used to frame interior walls and partitions. Aluminum is sometimes used for interior studs. Deck forms made of corrugated steel (*Figure 7*) are used to form concrete floors in commercial buildings. Steel reinforcing bars and mesh are embedded in concrete forms to strengthen poured concrete. Anyone drilling into concrete should recognize the possibility of hitting the rebars or steel deck forms.

103F05.EPS

Figure 5 Structural steel construction.

103F06.EPS

Figure 6 Structural steel framing.

Figure 7 Corrugated steel deck form.

2.2.1 Steel Framing Materials

Steel framing components include steel studs and, in some cases, steel joists and roof trusses (*Figure 8*). The vertical and horizontal framing members serve as structural load-carrying components for a variety of low- and high-rise structures. Steel stud framing is compatible with many types of surfacing materials.

The advantages of steel framing include fire resistance, uniformity of dimension, lightness of weight, freedom from rot, and ease of construction. The components of steel frame systems are made to fit together easily. A variety of steel framing systems are available in both loadbearing and nonbearing types. Some nonbearing partitions are designed to be demountable or moveable. These still meet the requirements of sound insulation and fire resistance when they are covered with the proper gypsum system.

When steel studs are used for drywall framing systems, the channel stock comes in two grades. It features knurled sides for positive screw settings. The first grade is the standard drywall stud (*Figure 9*). Standard studs come in widths of 1⅝" to 6". The flanges are 1⅜" × 1¼". Lengths of 6' to 16' are commercially available. The standard drywall stud is 25-gauge steel (the higher the gauge, the lighter the metal). Depending on the product number, lengths range from 8' to 12'. Other lengths are available by special order.

The second grade is the extra heavy drywall stud (*Figure 10*). These studs have knurled sides for positive screw settings. They also have cutouts and utility knockout holes 12" from each end and at the midpoint of the stud.

The width of the extra heavy studs varies from 1⅝" to 6". The flanges are 1⅜" × 1¼". The extra heavy drywall stud is 20 gauge. This type of stud can be ordered in any length that is needed.

Figure 8 Steel trusses.

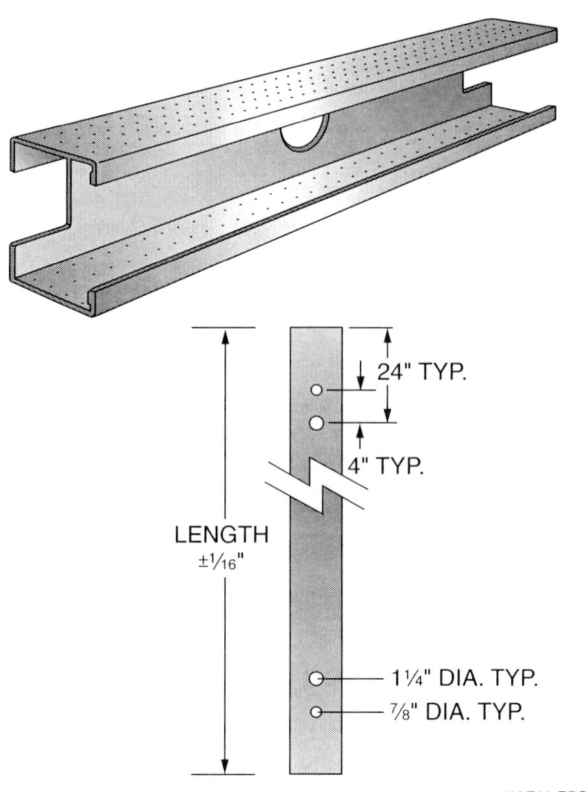

Figure 9 Standard steel stud stock.

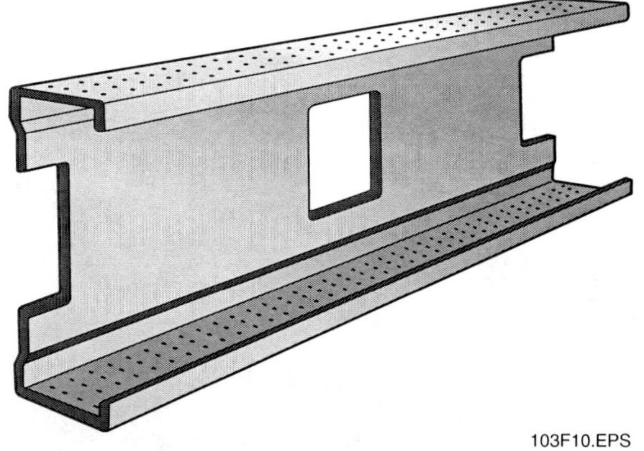

Figure 10 Heavy-duty stud stock.

Figure 11 Curtain wall under construction.

Inside Track

Identifying Structural Studs

Structural studs are marked with a color code for easy identification. The coding is as follows:

Gauge	Color
20	White
18	Yellow
16	Green
14	Blue
12	Red

Note that both light and structural gauge studs are made of 20-gauge steel. The difference is in the dimensions.

Steel studs are also available in greater strengths of gauges 18, 16, 12, and 10. These strengths are classified as structural steel studs. They are available in widths of 2½" to 8". The flanges are 1⅝" × 2" or 2½". They can be ordered in whatever length is needed.

3.0.0 COMMERCIAL CONSTRUCTION METHODS

The structural framework of large buildings, such as office buildings, hospitals, apartment complexes, and hotels, is usually made from concrete or structural steel. The exterior finish is often concrete panels that are either prefabricated and raised into place, or poured into forms built at the site. Floors are usually made of concrete that is poured at the site using wood, metal, or fiberglass forms. Exterior walls (curtain walls) may also be made of glass in a metal or concrete framework (*Figure 11*). A curtain wall section being installed is shown in *Figure 12*. Before the concrete is poured, provisions must be made for cabling pathways.

Figure 13 shows the structure of a building in which all of the structural framework is made of concrete poured at the site. Each component of the structure requires a different type of form. In this case, the floor and beams were made in a single pour using integrated floor and beam forms. The forms were removed once the concrete hardened.

Figure 12 Installing a corner curtain wall panel.

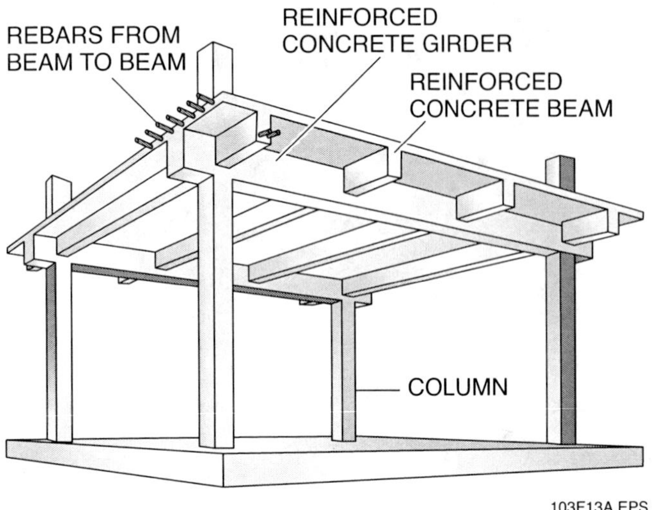

Figure 13   Structural concrete building.

In some commercial applications, tilt-up concrete construction is used. In tilt-up construction, the wall panels are usually poured on the concrete floor slab and then tilted into place on the footing using a crane (*Figure 14*). The panels are welded together.

The main difference between tilt-up and other types of large commercial construction is that there is no steel or concrete framework in tilt-up construction. The walls and floor slab bear the entire load. Tilt-up construction is most common in one- or two-story buildings that have a slab at grade (no

Inside Track

Masonry Curtain Wall

This masonry curtain wall panel combines 2"-thick architectural precast concrete brick with a heavy-gauge stainless steel frame and insulated stainless steel anchor doors.

Figure 14 Tilt-up panels being lifted into place.

under the forms to support them until the concrete hardens. *Figure 15* shows cellular floors poured over corrugated steel forms. The forms remain in place to provide channels for running cabling.

A section of metal, plastic, or fiber sleeve is often inserted vertically into the form before the concrete is poured. This allows for electrical, communications, and other cabling to pass through the floor.

In some installations, underfloor duct systems are embedded in the concrete floor. They are used to provide horizontal distribution of cables. Vertical access ports (handholes) are embedded in the form so that cable can be fished to various locations in the space. *Figure 16* shows a single-level feeder duct system. In two-level systems, one level carries electrical power cables and the other carries low-voltage cables.

Trench ducts are metal troughs that are embedded in the concrete floor. They are used as feeder ducts for electrical power and telecommunication lines (*Figure 17*).

Access floors consist of modular floor panels supported by pedestals. They may or may not have horizontal bracing in addition to the pedestals. This type of structure is used in computer rooms, intensive care facilities, and other areas where a large amount of cabling is required. In some applications, such as factories, a trench may be formed in the concrete floor to accommodate cabling and other services.

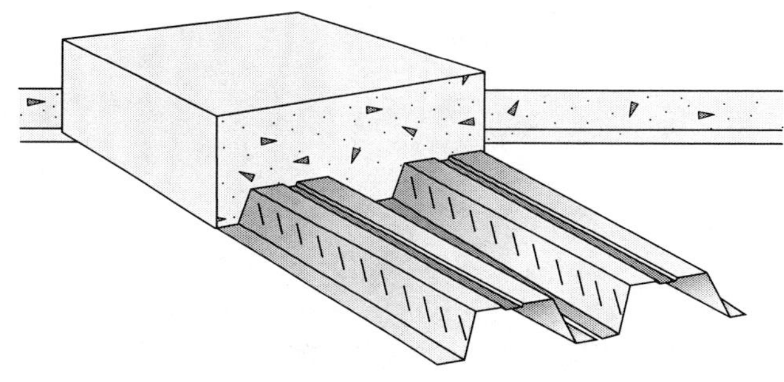

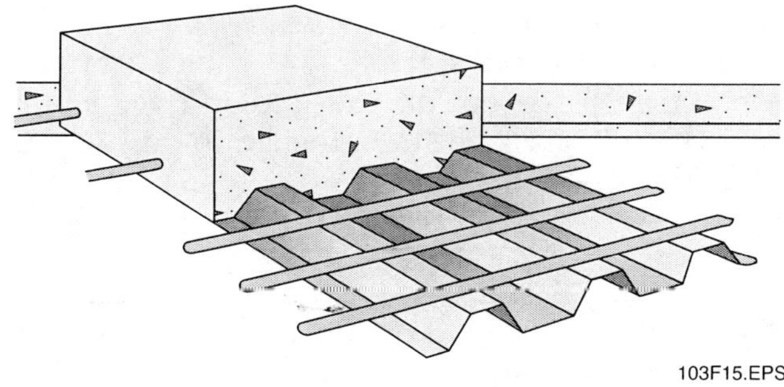

Figure 15 Corrugated steel forms.

below-grade foundation). It is popular for warehouses, low-rise offices, churches, and a variety of other commercial and multi-family residential applications. Tilt-up panels of 50' in height are common. Panels typically range from 5" to 8" thick, but thicker walls can be obtained when using lighter-weight concrete.

3.1.0 Floors

Once the framework is in place, the concrete floors are poured using deck forms. Shoring is placed

Concrete decks can be strengthened using post-tensioning cables rather than rebar. The steel cables are embedded in the concrete form (*Figure 18*) and protrude from the ends of the concrete beams (*Figure 19*). These cables can be 3" or more thick. When the concrete has hardened, hydraulic jacks are used to tension the cables. The cables are then anchored in place, cut off, and sealed (*Figure 20*). Concrete that has been strengthened in this way is called post-tensioned concrete.

3.2.0 Exterior Walls

When walls are formed of concrete, openings for doors and windows are made by inserting wooden or metal bucks in the form, as shown in *Figure 21*. Openings for services, such as piping and cabling, are made by inserting fiber, plastic, or metal tubes into the form.

3.3.0 Interior Walls and Partitions

Although they are sometimes used in residential construction, steel studs are the standard for framing walls and partitions in commercial construction. After the studs have been installed, one or more layers of gypsum wallboard and insulation are applied. The type and thickness of the wallboard and insulation depend on the fire rating and soundproofing requirements. Soundproofing needs can vary. They are often based on the amount of privacy required. For example, executive and medical offices may require more privacy than general-use offices.

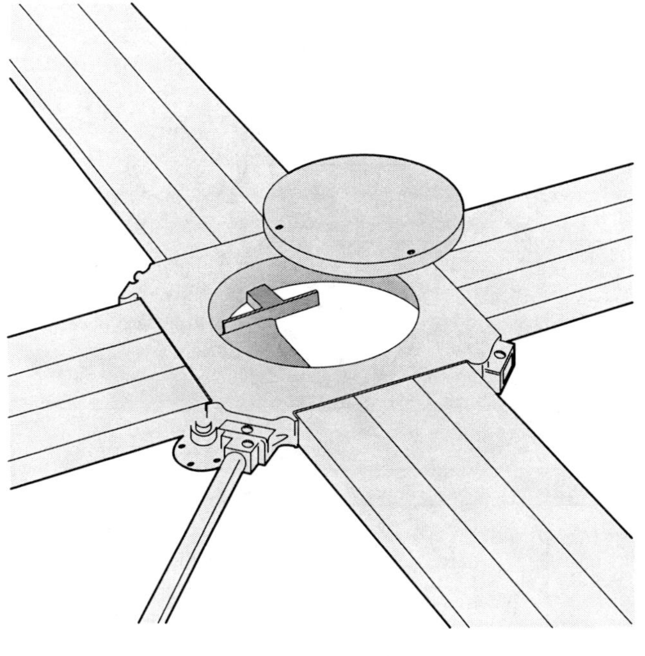

Figure 16   Flushduct underfloor system.

The fire rating specified by the building code determines the types and amount of material used in a wall or partition. For example, a one-hour rated wall might be made of single sheets of ⅝" gypsum wallboard on wooden or 25-gauge metal studs. A two-hour rated wall would require heavy-gauge metal studs and two layers of fire-resistant gypsum wallboard. Fire ratings are discussed in more detail later in this module.

3.3.1 Framing Materials

The general approach to framing with metal studs is the same as that used for wooden studs. In fact, metal studs can be used with either wood plates or metal runners.

Like wood framing, metal framing is installed 12", 16", or 24" on center. Openings are framed with headers and cripples. Special framing is needed for corners and partition Ts.

Depending on the load, reinforcement may be needed when framing openings. Bracing of walls to keep them square and plumb is also required.

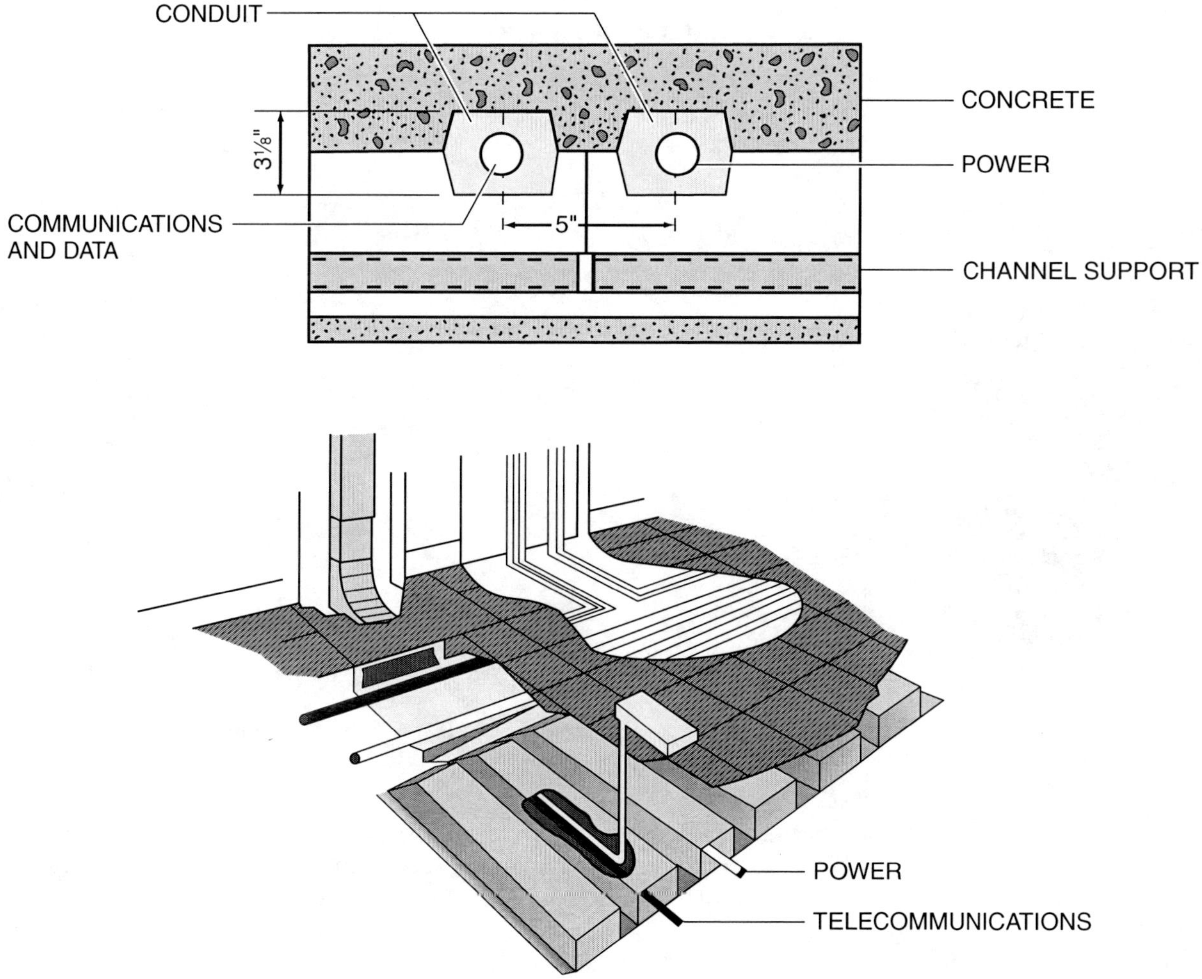

Figure 17   Trench duct in a cellular floor.

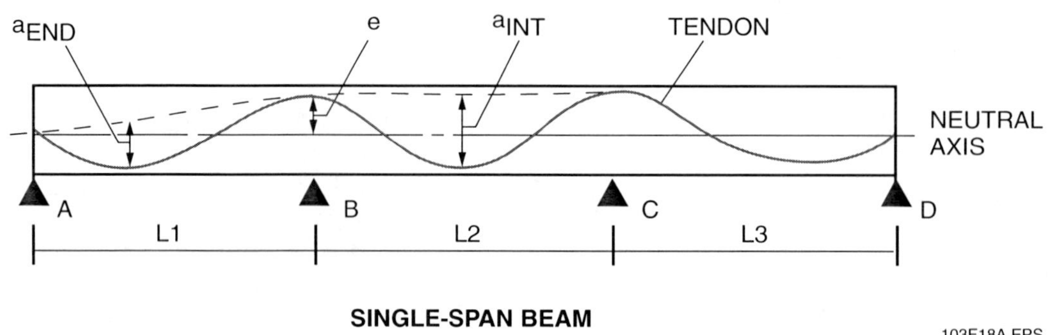

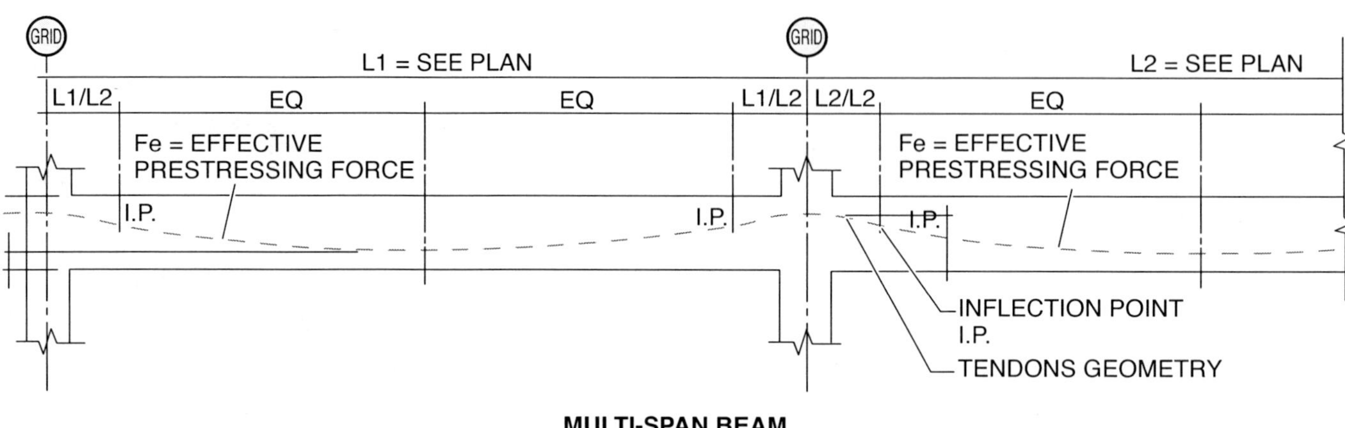

Figure 18 Examples of tendon profiles.

Figure 19 Post-tensioning tendons.

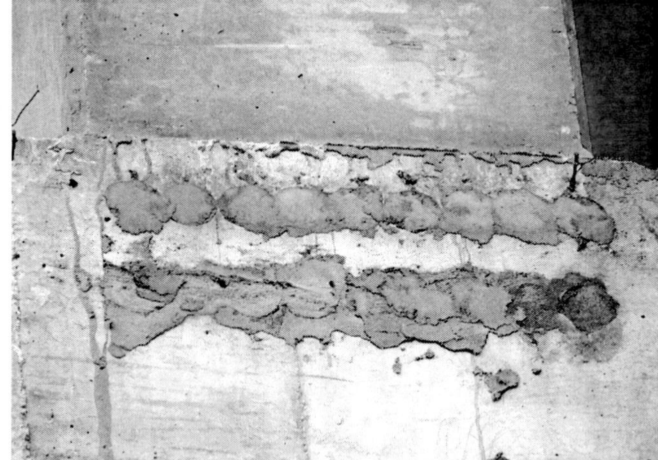

Figure 20 Sealed tendon openings.

The illustrations in this section show examples of common framing techniques. *Table 1* shows the framing spacing for various gypsum drywall applications.

The installation of metal studs typically starts by laying metal tracks in position on the floor and ceiling and securing them (*Figure 22*). When the tracks are being applied to concrete (*Figure 23*), a low-velocity, powder-actuated fastener is generally used. When the tracks are being applied to wood joists, such as in a residence, screws can be driven with a screw gun.

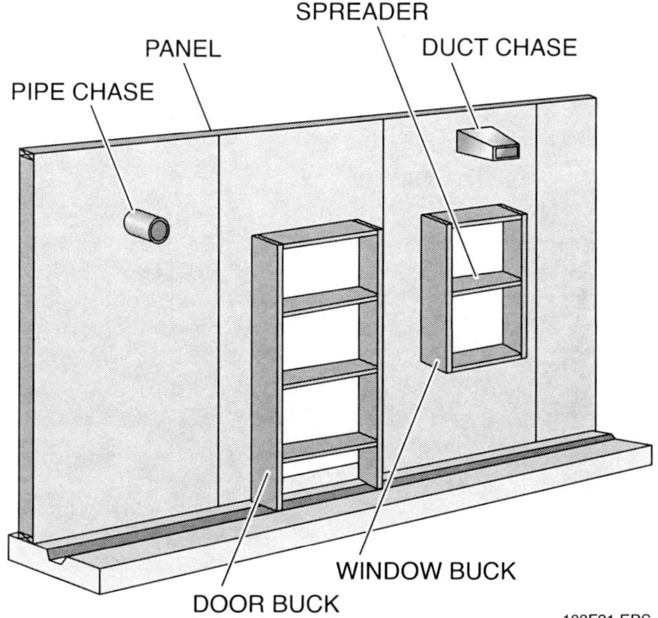

Figure 21   Framing openings in concrete walls.

Once the tracks are in place, the studs and openings are laid out in the same way as a wood frame wall. The studs may be secured to the tracks with screws, or they may be welded. In some cases, the entire wall is laid out on the floor and then raised and secured. When heavy-gauge walls are used, they may be assembled and welded in the shop and brought to the site.

There are some differences between installing steel and wooden nonbearing partitions. When building with wood, all partitions must be nailed together. With steel studs, this is not required.

As shown in *Figure 24*, the partitions are held back from the other partitions so that the drywall will slide past. Note the conduit fed through the openings in the stud.

Table 1   Maximum Framing Spacing

	Single-Ply Gypsum Board Thicknesses	Application to Framing	Maximum OC Spacing of Framing
Ceilings	⅜"	Perpendicular	16"
	½"	Perpendicular or Parallel	16"
	½"	Perpendicular	24"
	⅝"	Perpendicular	24"
Sidewalls	⅜"	Perpendicular or Parallel	16"
	½" or ⅝"	Perpendicular or Parallel	24"

Fasteners Only – No Adhesive Between Plies

	Multi-Ply Gypsum Board Thicknesses		Application to Framing		Maximum OC Spacing of Framing
	Base	Face	Base	Face	
Ceilings	⅜"	⅜"	Perpendicular	Perpendicular	16"
	½"	⅜"	Parallel	Perpendicular	16"
		½"	Parallel	Perpendicular	16"
	½"	½"	Perpendicular	Perpendicular	24"
	⅝"	½"	Perpendicular	Perpendicular	24"
		⅝"	Perpendicular	Perpendicular	24"

Sidewalls*

*For two-layer applications with no adhesive between plies, ⅜", ½", or ⅝" gypsum board may be applied perpendicularly (horizontally) or parallel (vertically) on framing spaced a maximum of 24" OC. Maximum spacing should be 16" OC when ⅜" board is used as the face layer.

103T01.EPS

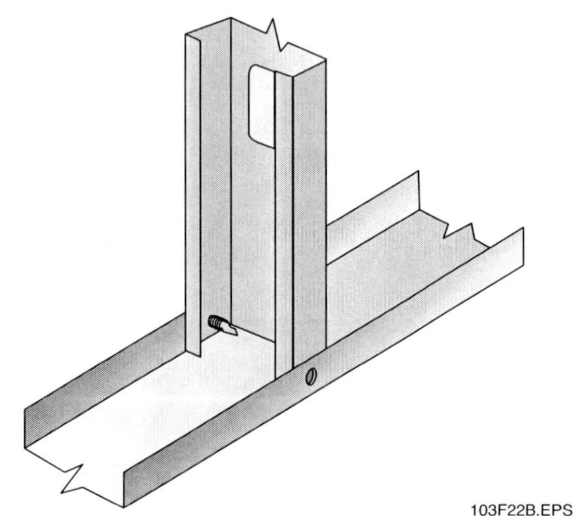

Figure 22 Steel framing.

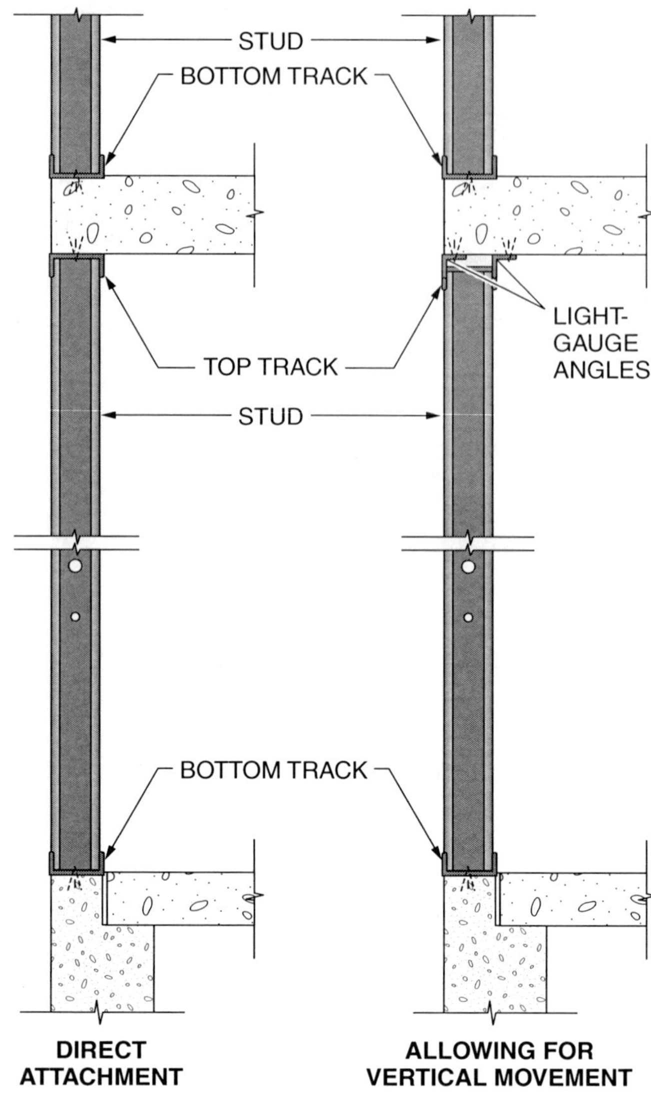

Figure 23 Steel studs with concrete floors and ceiling.

When metal studs are used to frame around steel beams, the metal studs are secured to the beam with powder-actuated fasteners, if allowed (*Figure 25*). The support members are screwed to the metal studs.

When the steel studs are installed against metal channels or flanges, the studs are secured to the channel with scrap pieces of metal studs (*Figure 26*).

In-line framing, or direct alignment, is the preferred and most common framing method. It provides a direct path for transferring the load from studs to joists, through the framing system, to the ground. In this method, cold-formed steel framing members are aligned vertically so that the center line of the joist web is within ¾" of the center line of the structural stud member below, or the center line of the stud web is within ¾" of the center line of the web joist below (*Figure 26*).

Figure 24 Partition held back to allow drywall to slide by.

Figure 25 Powder-actuated tool in use.

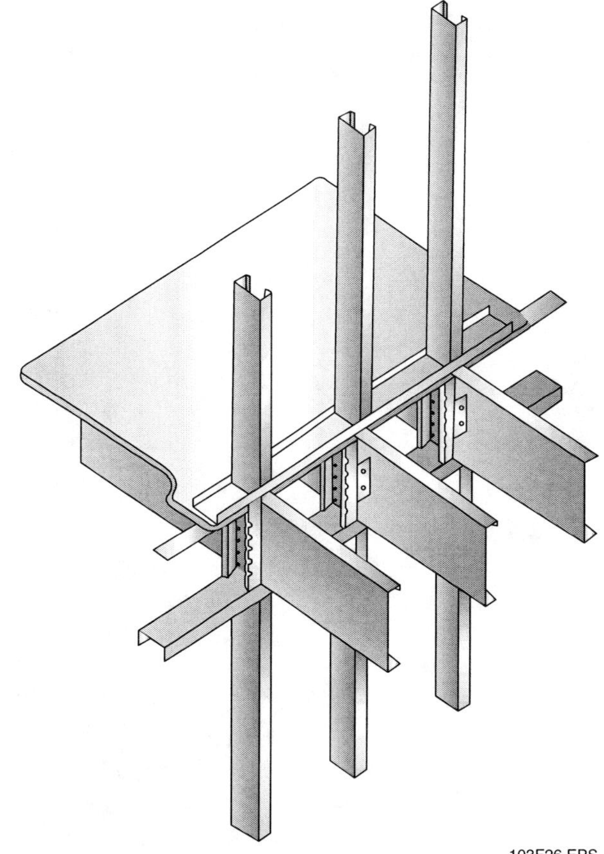

Figure 26 Studs secured to a channel.

3.3.2 Bracing Walls

Different types of bracing are used to support metal stud walls. Lateral bracing consisting of continuous metal strapping is always recommended as the minimum support for metal stud walls (*Figure 27*). Diagonal bracing that uses metal strapping is sometimes required (*Figure 28*). This is done with 20-gauge, 2" wide metal straps placed close to the end of the wall. Lateral and diagonal braces can be screwed or welded to the studs.

For heavier studs (6" and wider), steel channel is threaded through the openings in the studs and welded to angle clips (*Figure 29*). Fine-gauge lateral bracing, fed through the openings in the studs and welded to each stud, is also used in some applications (*Figure 30*).

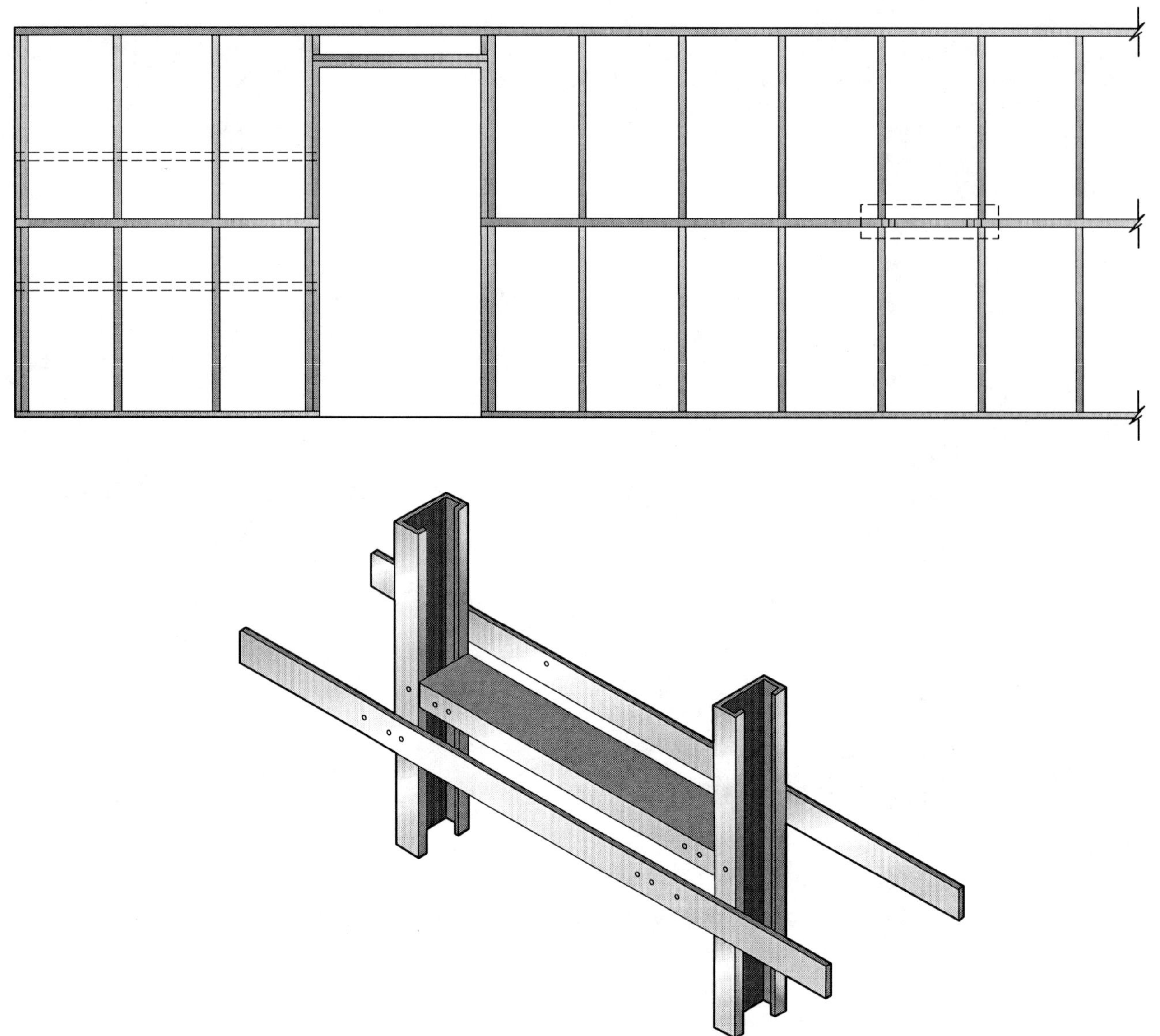

Figure 27   Lateral strapping for stud walls.

3.3.3 Door and Window Openings

The door and window openings in steel-framed structures are similar to those found in wood-framed structures. Loadbearing headers, such as the one shown in *Figure 31*, combine C-shapes with cripple studs to provide the required capacity. Headers in nonbearing walls are of simpler construction (*Figure 32*). Steel girders may be used as headers for large openings in loadbearing walls.

The number of studs in the jamb will vary depending on the load and the engineering requirements of the project. As shown in *Figure 33*, there could be several studs.

Figure 28   Diagonal strapping for shear walls.

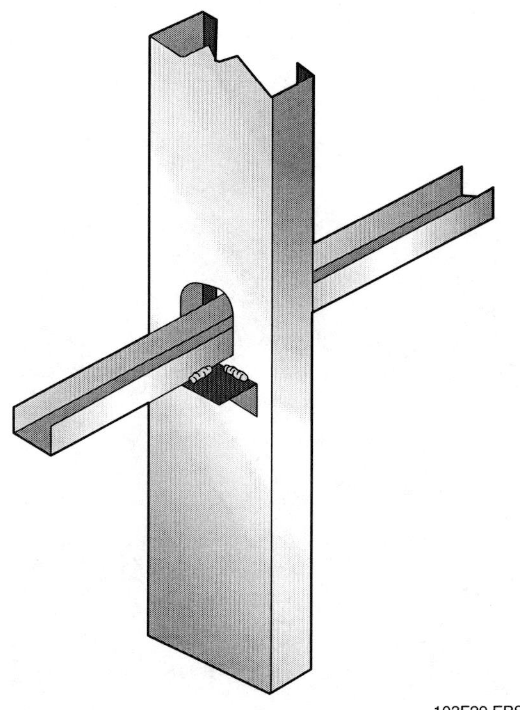

Figure 29   Heavy-duty bracing.

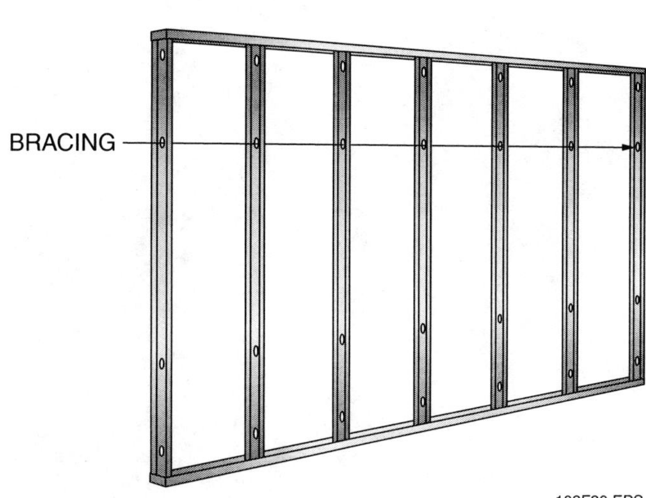

Figure 30   Fine-gauge lateral bracing.

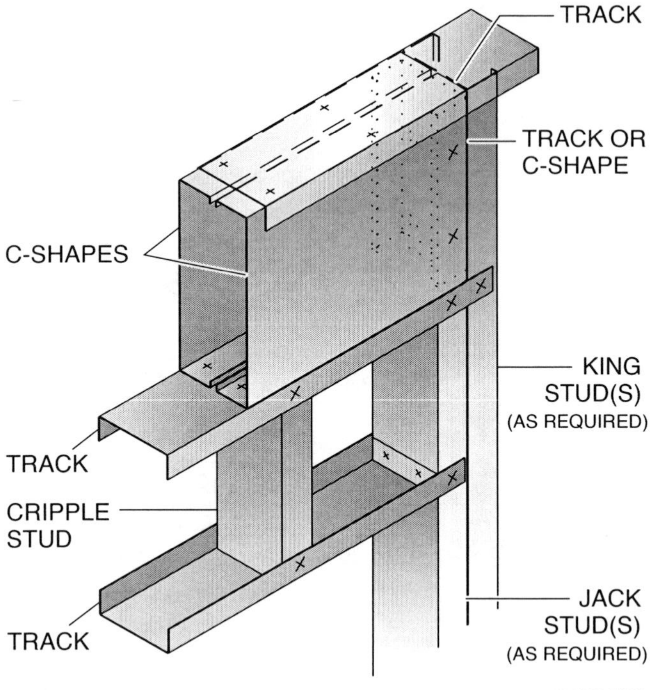

Figure 31 Loadbearing header.

Figure 32 L-header.

Figure 33 Rough opening.

3.4.0 Roof Structure

In commercial construction, roofs are generally supported by open-web steel joists (trusses) (*Figure 34*). These joists can be flat or sloped. Flat joists are used to support flat roofs, as well as decks in multi-story buildings. A truss may span anywhere from 20' to 80'. The open-web design allows for installation of piping, ducts, and conduit.

3.5.0 Ceilings

Suspended ceilings are sometimes found in residential applications. However, they are most often used in commercial construction. Suspended ceilings have a number of advantages in commercial work:

- They offer excellent noise suppression.
- They provide an area in which horizontal runs of cabling, piping, heating and cooling ducts, and other services can be readily accessed.

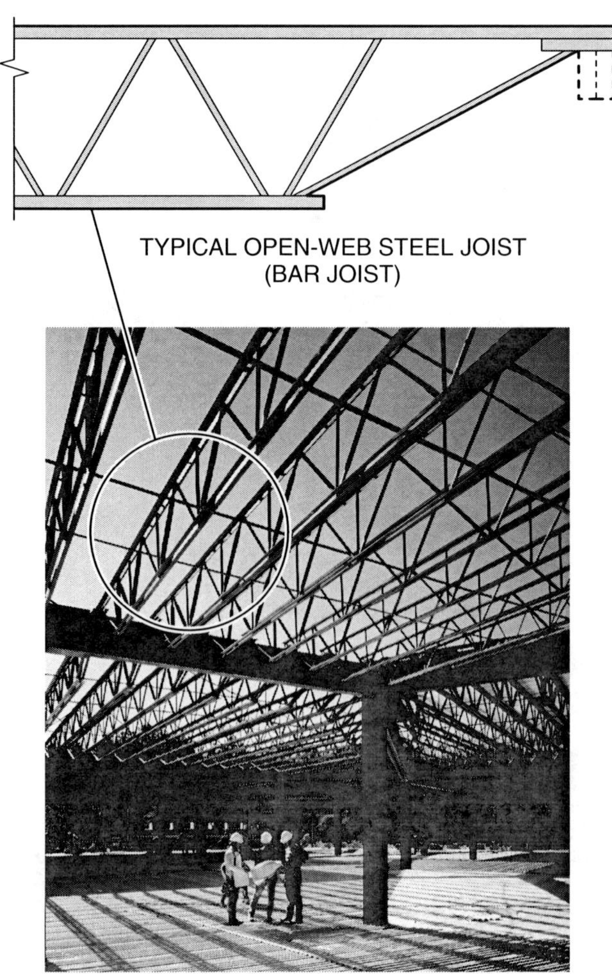

Figure 34 Open-web steel joists.

- In many commercial buildings, the area between the suspended ceiling and the floor above acts as the return air plenum for air conditioning and heating. This eliminates the need for some of the sheet metal ductwork.
- The use of suspended ceilings eliminates the need for ceiling framing and the need to box in horizontal runs of ductwork and piping.

A wide variety of suspended acoustical ceiling systems is available. They all use the same basic materials, but their appearances can be completely different. The focus of this module is on the following systems:

- Exposed grid
- Metal pan
- Direct-hung concealed grid
- Integrated ceiling
- Luminous ceiling
- Suspended drywall furring ceiling

3.5.1 Exposed Grid Systems

In an exposed grid suspended ceiling, a light metal grid is hung by wire from the original ceiling or the deck above. This type is also called a direct-hung system. Ceiling panels, which are usually 2' × 2' or 2' × 4', are then placed in the frames of the metal grid. Exposed grid systems are built using the components and materials shown in *Figure 35*.

- *Main runners* – These are the primary support members of the grid system for all types of suspended ceiling systems. Main runners are 12' in length. They are usually made in the form of an inverted T. When it is necessary to lengthen the main runners, they are usually spliced together using extension inserts. However, the method of splicing the main runners varies with the type of system being used.

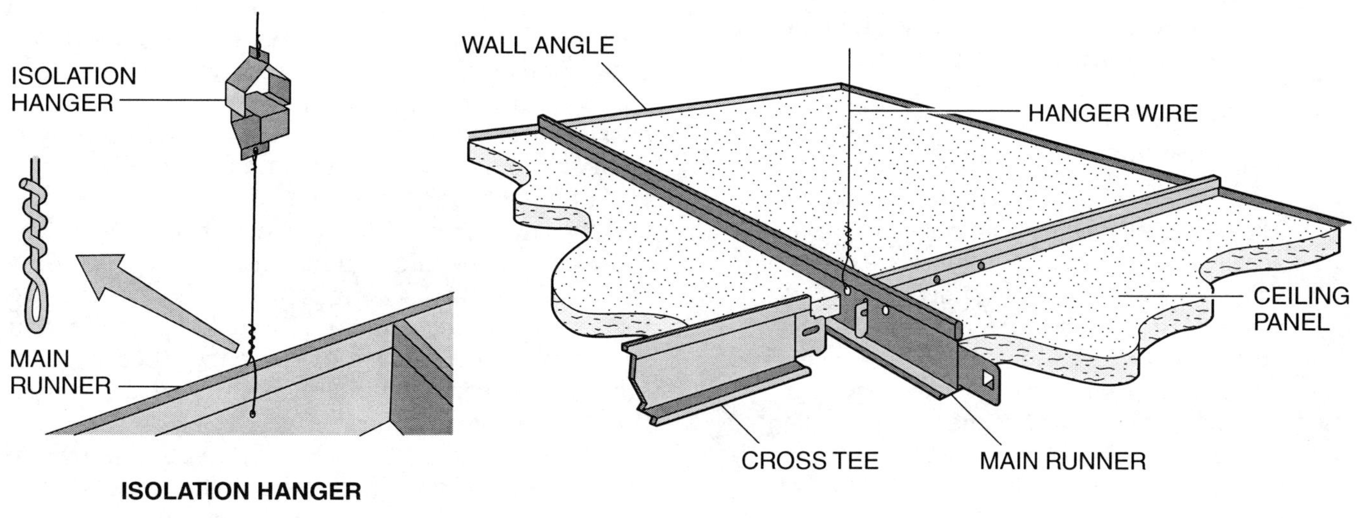

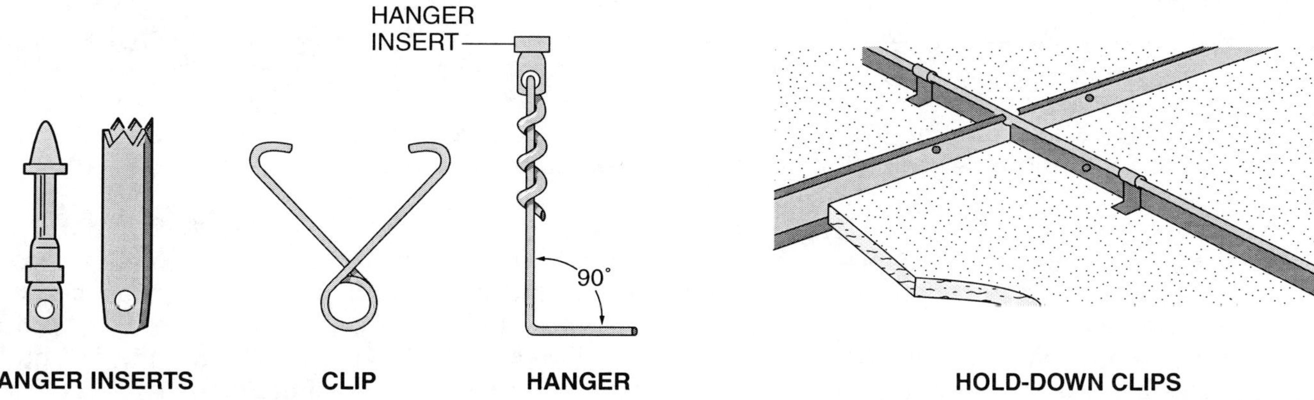

Figure 35 Typical exposed grid system components.

- *Cross runners (cross ties or cross tees)* – These supports are inserted into the main runners at right angles. They are spaced an equal distance from each other, forming a complete grid system. Cross runners are held in place either by clips or automatic locking devices. Typically, they are 4' or 2' in length. They are usually made in the form of an inverted T. Note that 2' cross runners are only required for use with 2' × 2' ceiling panels.
- *Wall angle* – These supports are installed on the walls to support the exposed grid system at the outer edges.
- *Ceiling panels* – The panels are laid in place between the main runners and cross ties to provide sound control. Acoustical panels used in suspended ceilings absorb sound waves. The panels are designed with many tiny sound traps consisting of drilled or punched holes and/or openings. A wide variety of ceiling panel designs, patterns, colors, facings, and sizes are available. Panels are usually made of glass or mineral fiber. In general, glass panels absorb more sound than mineral fiber panels. Panel facings are often made of embossed vinyl. They are available in a variety of patterns, such as fissured, pebbled, and striated. The ceiling panels used must be compatible with the ceiling suspension system. Not all panels fit all systems.

<table><tr><td>NOTE</td><td>The terms *ceiling panel* and *ceiling tile* have specific meanings. Ceiling panels are any lay-in acoustical board that is designed for use with an exposed mounting system. Ceiling panels do not have finished edges or precise dimension tolerances because the exposed grid system support members provide the trimout. Ceiling tiles are acoustical ceiling boards, usually 12" × 12" or 12" × 24". They are nailed to, cemented to, or suspended by a concealed grid system. The edges are often notched and cut back.</td></tr></table>

- *Hanger inserts and clips* – Many types of fasteners are used to attach the grid system hangers or wires to the building's horizontal structure above the suspended ceiling. Screw eyes and star anchors are commonly used. These require an electric hammer for installation. Eye pin fasteners are also used to fasten into reinforced concrete with a powder-actuated fastening tool. Clips are used when beams are available. Clips are typically installed over the beam flanges.

The hanger wires are then inserted through the loops in the clips and secured. These devices must be able to handle the load.

- *Hangers* – Hangers are attached to the hanger inserts, pins, and clips to support the suspended ceiling's main runners. The hangers can be made of No. 12 wire or heavier rod stock. Ceiling isolation hangers are also available to isolate ceilings from the noise traveling through the building's structure.
- *Hold-down clips* – These clips are used in some systems to hold the ceiling panels in place.
- *Nails, screws, rawls, toppets, and molly bolts* – These fasteners are used to secure the wall angle to the wall. The specific item used depends on the wall construction and material.

3.5.2 Metal Pan Systems

The metal pan system is similar to the conventional suspended acoustical ceiling system, but metal tiles or pans are used in place of the conventional acoustical panels (*Figure 36*).

The pans are made of steel or aluminum and are generally painted white, although other colors are available by special order. Pans are available in a variety of surface patterns. Tests have shown that metal pan ceiling systems are effective for sound absorption. They are durable and easily cleaned and disinfected. In addition, the finished ceiling is unlikely to sag. The metal pans are die-stamped and have crimped edges that snap into the spring-locking main runner to provide a flush ceiling.

Care must be taken in handling the pans if they must be removed. Using white gloves or rubbing the hands with cornstarch will keep any sweat stains from the surface of the pans. If care is not taken, fingerprints will be plainly visible when the units are reinstalled.

When removing or installing pans, grasp the pan at its edge and force its crimp into the tee bar slots. Use the palms of your hands to seat the pan. After installing several of the pans, slide them along the tee bars into position. Use the side of your closed fist to bump the pan into level position if it does not seat readily. If metal pan hoods are required, slip them into position over the pans as they are installed. The purpose of the hoods is to reduce the sound travel through the ceiling into the room.

If a metal pan must be removed, a pan pulling device can be used (*Figure 37*). To pull out a pan, insert the tool into the joint between the pans. Push the tool up until it engages the pan, and then gently pull down. Using this removal procedure will reduce the danger of bending the pan out of shape. A metal pan ceiling is shown in *Figure 38*.

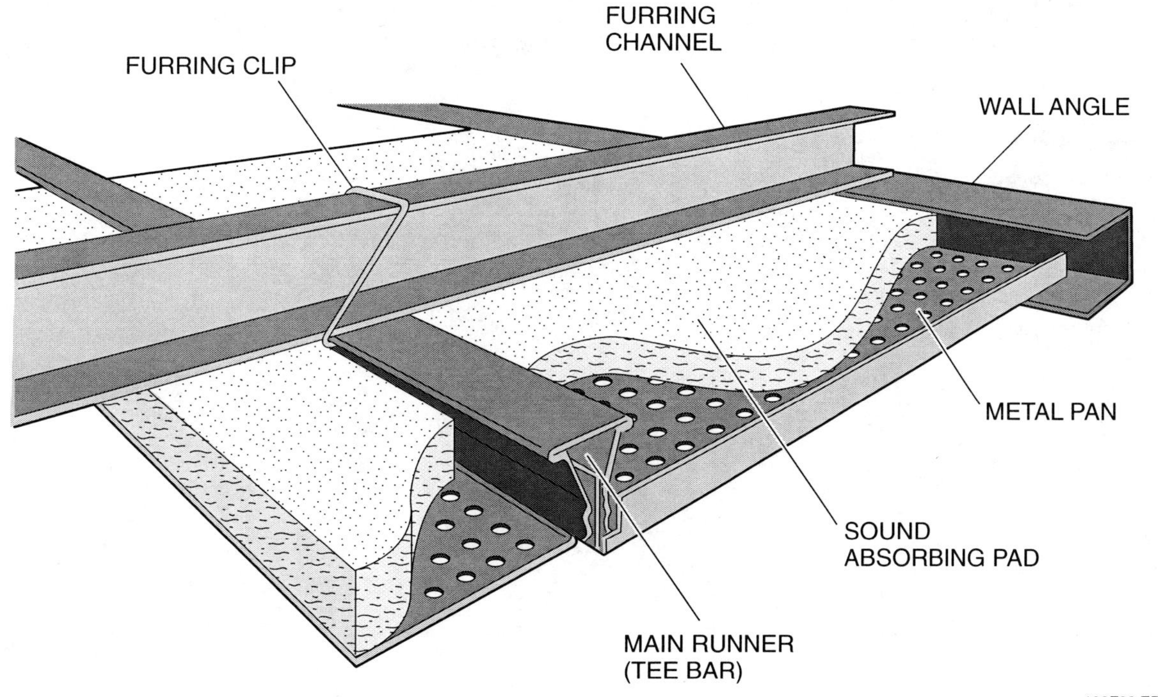

Figure 36 Metal pan ceiling components.

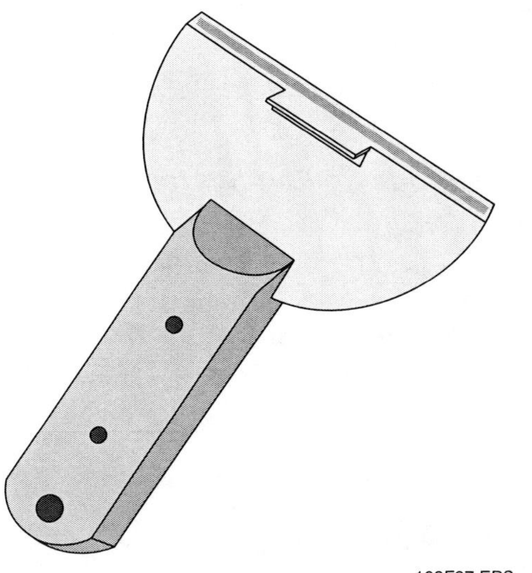

Figure 37 Pan removal tool.

Figure 38 Metal pan ceiling.

3.5.3 Direct-Hung Concealed Grid Systems

In direct-hung concealed grid systems, the support runners are hidden from view. This results in a patterned ceiling that is not broken by the pattern of the runners (*Figure 39*).

The tiles used for this system are similar to common acoustical tile, but are manufactured with a kerf on all four edges. Kerfed and rabbeted 12" × 12" or 12" × 24" tiles are used with this system. Tiles of various colors and finishes are available. For a diagram of a concealed grid system installation, see *Figure 40*.

If regular access is needed to the area above the ceiling, special access systems can be incorporated into the ceiling (*Figure 41*).

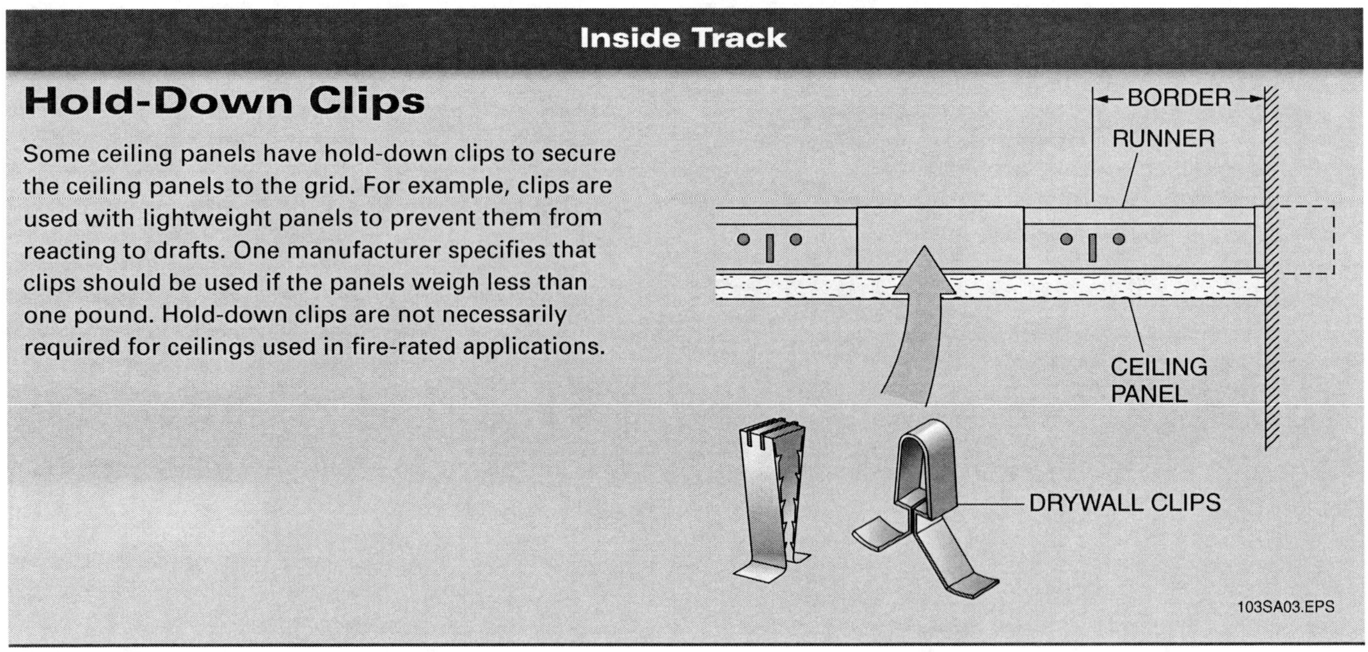

Figure 39   Concealed grid system.

3.5.4 Integrated Ceiling Systems

Integrated ceiling systems include the lighting and/or air supply diffusers as part of the overall ceiling system, as shown in *Figures 42 and 43*.

These systems are available in units called modules. The common sizes are 30" × 60" and 60" × 60". The dimensions refer to the spacing of the main runners and cross tees.

3.5.5 Luminous Ceiling Systems

Luminous ceiling systems (*Figure 44*) are available in many styles. Examples are an exposed-grid system with drop-in plastic light diffusers and an aluminum or wood framework with translucent acrylic light diffusers.

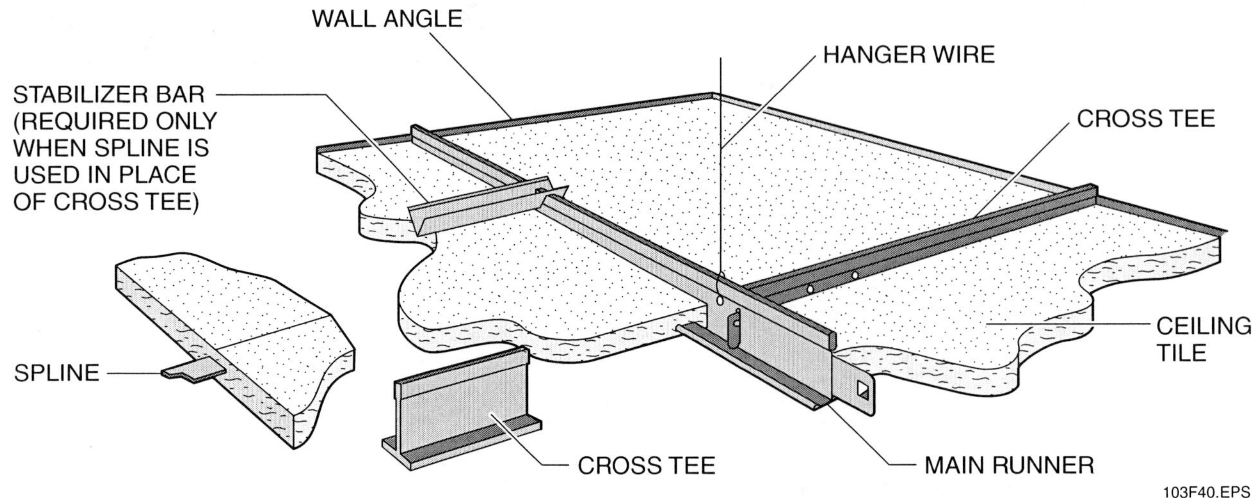

Figure 40   Direct-hung concealed grid system components.

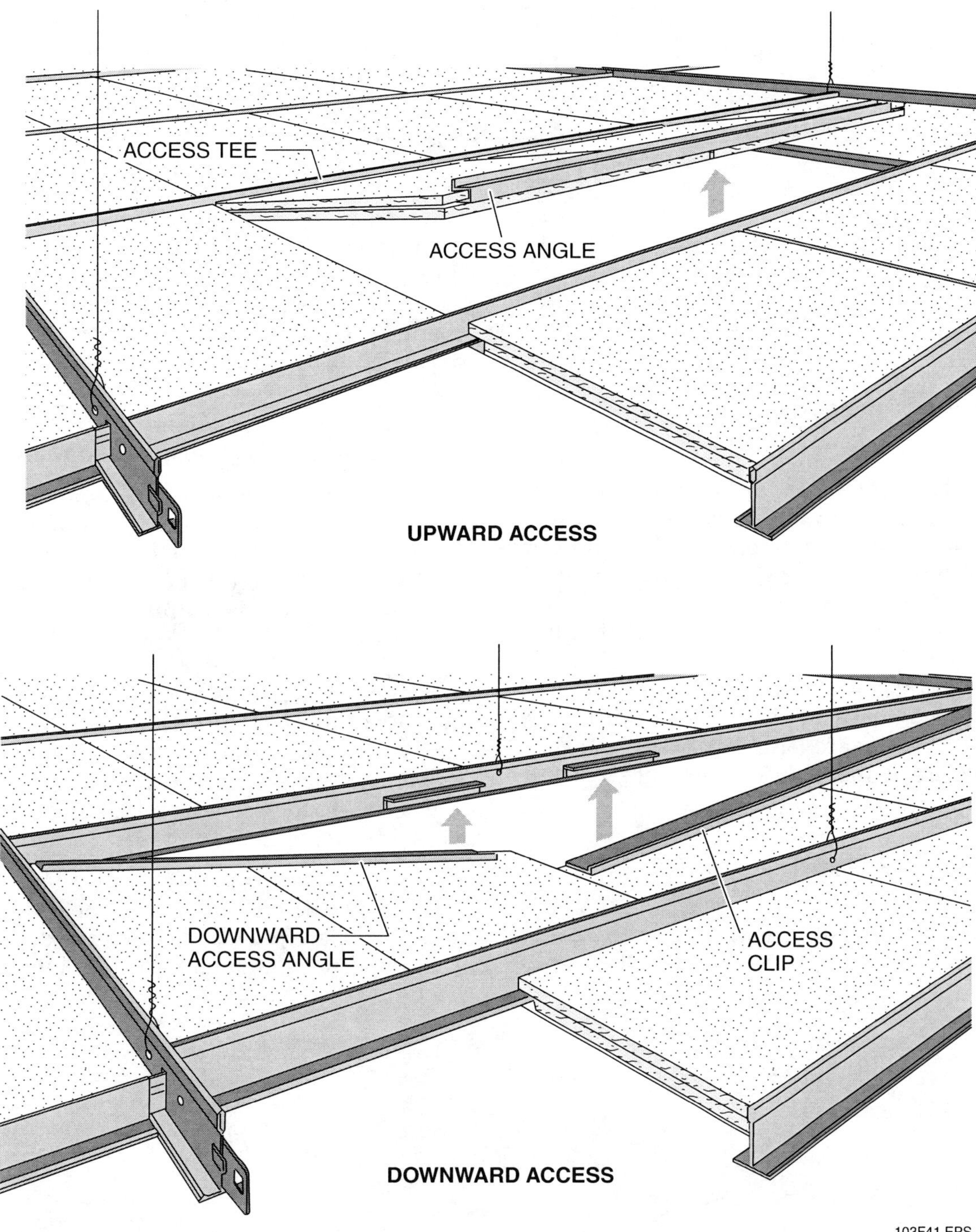

Figure 41 Typical access for concealed grid ceilings.

Fluorescent fixtures are generally installed above the translucent diffusers. Standard module sizes of 2' × 2' up to 5' × 5' are available, as are custom sizes for special applications. There are two types of luminous ceilings—standard and nonstandard. Standard systems are those that are available in a series of standard sizes and patterns. Nonstandard systems deviate from the normal spacing of main supports and can have unusual sizes, shapes, and configurations in their diffusing panels.

All surfaces in the luminous space, including pipes, ductwork, ceilings, and walls, are painted with a 75 to 90 percent reflectance matte white finish. Any surfaces in this area that might tend to flake, such as fireproofing and insulation, should receive an approved hard-surface coating prior to painting to prevent flaking onto the ceiling below.

Figure 42 Integrated grid system.

The installation of a standard luminous system is the same as that used for an exposed grid suspended system, with the exception of the border cuts. Luminous ceilings are placed into the grid members in full modules. Any remaining modules are filled in with acoustical material that has been cut to size.

In a 2' × 2' or 2' × 4' standard exposed grid system, luminous panels provide the light-diffusing element. The panels are laid in between the runners. Panels are available in a variety of sizes and shapes.

3.5.6 Suspended Drywall Furring Ceiling Systems

The suspended drywall furring ceiling system is selected when a drywall finish or drywall backing is required for an acoustical tile ceiling.

When this type of ceiling is installed, the first step is to install a carrying channel (*Figure 45*). Furring channels are then installed at right angles to the carrying channels. *Figure 46* shows a hat-type metal furring channel being used.

After the furring channels are in place, the drywall sheets are installed with drywall screws driven into the furring channel (*Figure 47*).

In some cases, the furring channels are attached directly to structural members, such as steel beams or wooden joists, instead of to suspended carrying channels. Ceiling tiles may also be attached to the drywall.

3.5.7 General Guidelines for Accessing Suspended Ceilings

In new construction, cable can sometimes be run before the suspended ceiling grid and panels are installed. More often, however, the cable installer is faced with the retrofit of an existing system or with a ceiling that has already been installed in a new building.

Keep in mind that ceiling panels are delicate and that grid systems are not capable of supporting much weight. In addition, many different types of ceiling systems are available, each with its own special requirements.

The following are general guidelines for working with suspended ceilings:

- Contact building maintenance personnel to find out how the ceiling is constructed and how to obtain spare panels in case some of the existing panels get damaged. The maintenance staff should also have the special tools you will need

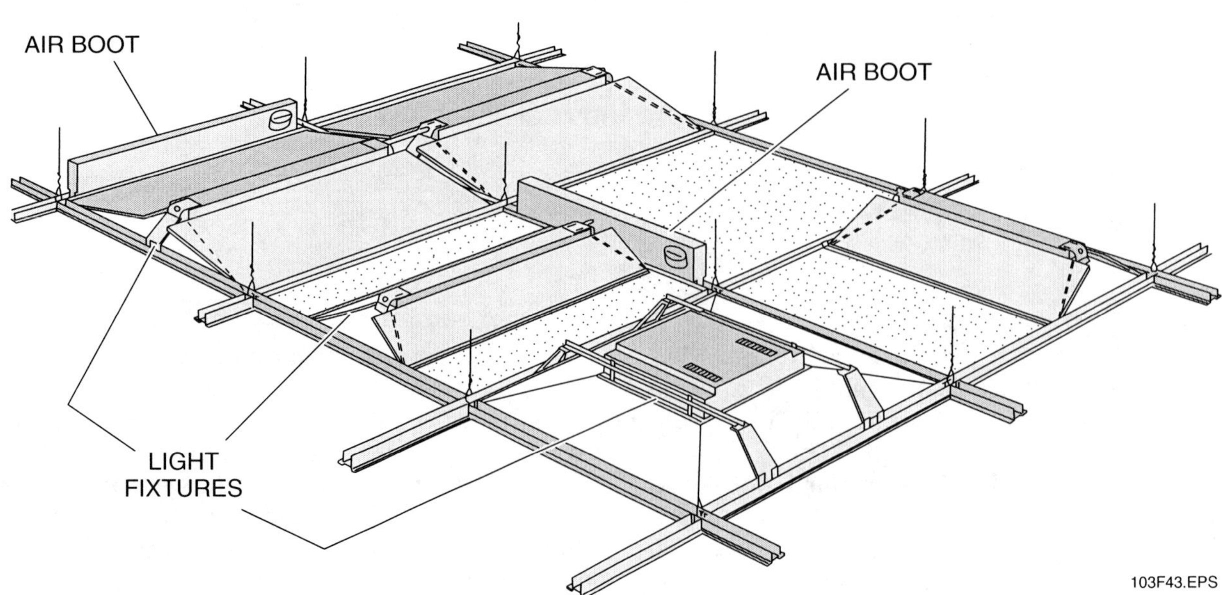

Figure 43 Integrated ceiling layout.

Figure 44   Luminous ceiling system.

to get access to some types of ceilings. For example, one type of concealed grid system has a special hook that is used to reach under the panel and release it from the cross member. As previously indicated, pan ceilings require special procedures for removing and installing pans.

- Never force ceiling panels. Some panels are clipped to the gridwork. If that is the case, find the panel that is not clipped and start there. A special tool may be needed to release the clips.
- Keep in mind that some ceilings have hinged panels that can be raised or lowered to provide access to the area above.

Inside Track

Plenum Ceilings

The systems that provide heating and cooling for most commercial buildings are forced-air systems. Blower fans are used to circulate the air. The blower draws air from the space to be conditioned and then forces it over a heat exchanger, which cools or heats the air. In a cooling system, the air is forced over an evaporator coil that has very cold refrigerant flowing through it. The heat in the air is transferred to the refrigerant, and the air that comes out the other side of the evaporator coil is cold. In homes, the air is delivered to the conditioned space and returned to the air conditioning/heating system through ductwork that is usually made of sheet metal. In commercial buildings with suspended ceilings, the space between the ceiling and the overhead decking is often used as the return air plenum. (A plenum is a sealed chamber at the inlet or outlet of an air handler.) This approach saves money by eliminating about half of the cost of materials and labor associated with ductwork.

One thing to keep in mind is that anything in the plenum space (electrical or data cable, for example) must be rated for plenum use in order to meet fire ratings. Plastic sheathing used on standard cables gives off toxic fumes when burned. Plenum-rated cable uses nontoxic sheathing.

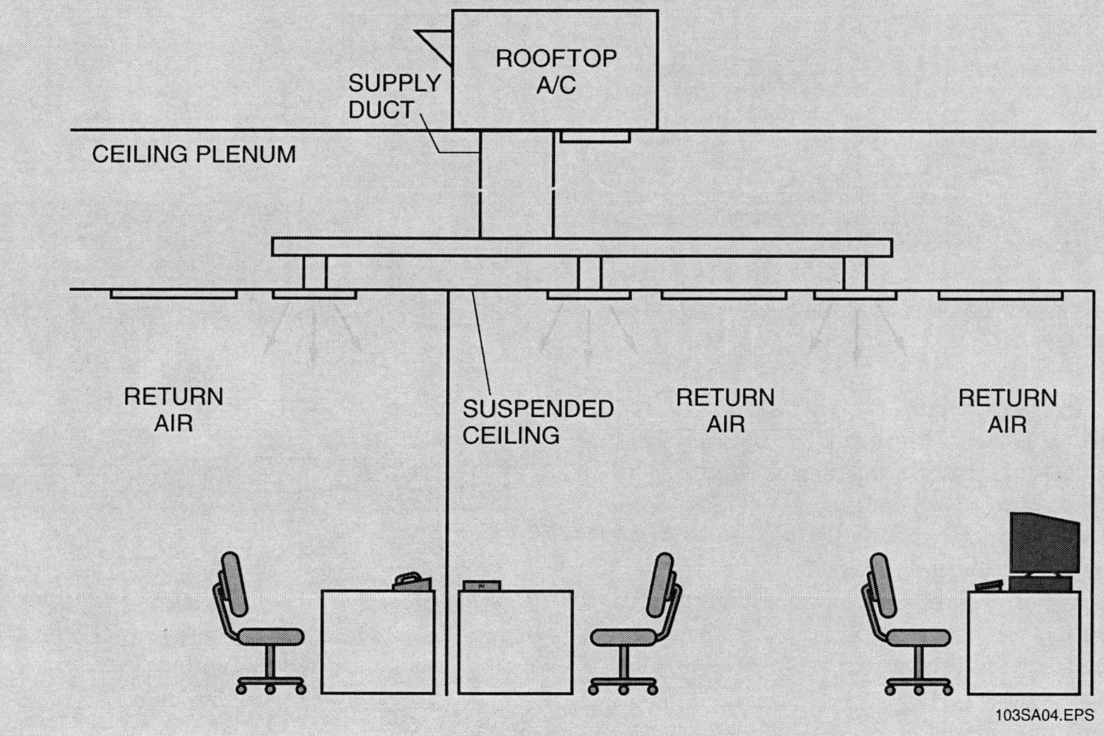

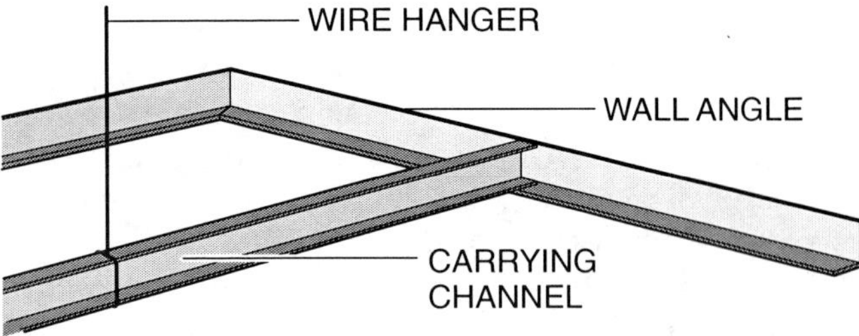

Figure 45   Carrying channel installed.

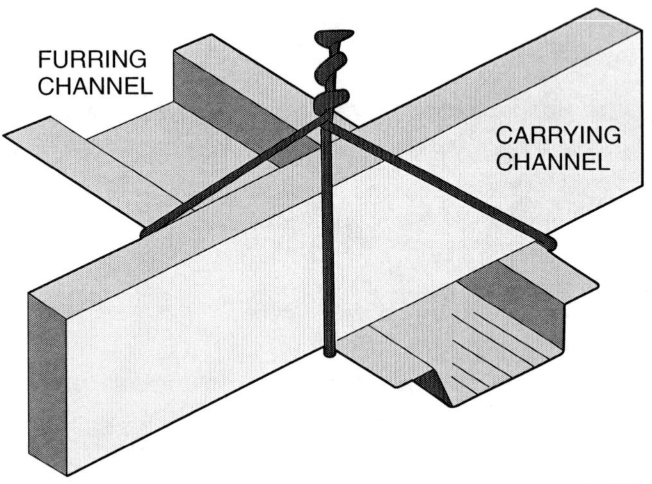

Figure 46   Furring channel attached to carrying channel.

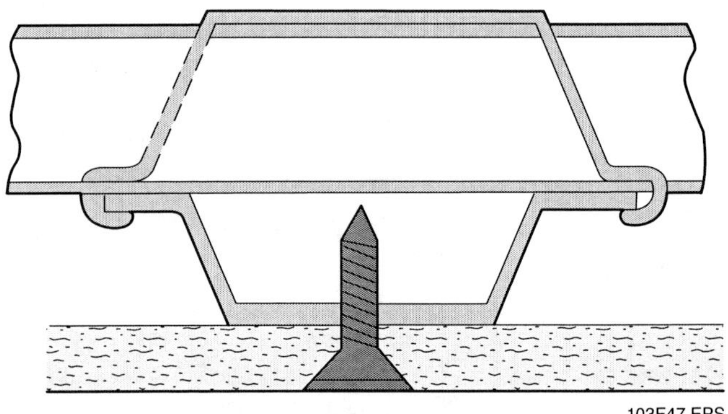

Figure 47   Drywall secured to furring channel.

- Make sure that your hands are clean in order to avoid staining the ceiling panels. If a panel gets dirty, try cleaning it with a damp sponge or an art gum eraser. Vinyl-faced fiberglass and mylar-faced ceilings can be cleaned with mild detergents or germicidal cleansers.
- Pan ceiling panels require special handling. Wear gloves or rub cornstarch on your hands to prevent the transfer of fingerprints to the panels.

4.0.0 FIRE-RATED AND SOUND-RATED WALLS

The construction of walls and partitions is regulated by the fire and soundproofing requirements specified in local building codes. In some cases, a frame wall with ½" gypsum drywall on either side is satisfactory. Extreme cases, such as the separation between the offices and the manufacturing spaces in a factory, may have special requirements. For example, it may be necessary to have a concrete block wall, combined with fire-resistant gypsum board, along with rigid and/or fiberglass insulation (*Figure 48*). This is especially true if there is any risk of explosion or fire.

While infrequently used in residential construction, steel studs are the standard for framing walls and partitions in commercial buildings.

Once the studs have been installed, one or more layers of gypsum board and insulation are applied. The type and thickness of the wallboard and insulation depend on the fire rating and soundproofing requirements. Soundproofing needs can vary and are often based on the amount of privacy required. For example, executive and medical offices may require more privacy than general offices.

The requirements for sound reduction and fire resistance can have a major effect on the thickness of a wall. A steel stud wall that has a high sound transmission class (STC) and fire resistance could have a total thickness of nearly 6". A low-rated wall could have a thickness of only 3", using steel studs, and 2¼", using wooden studs.

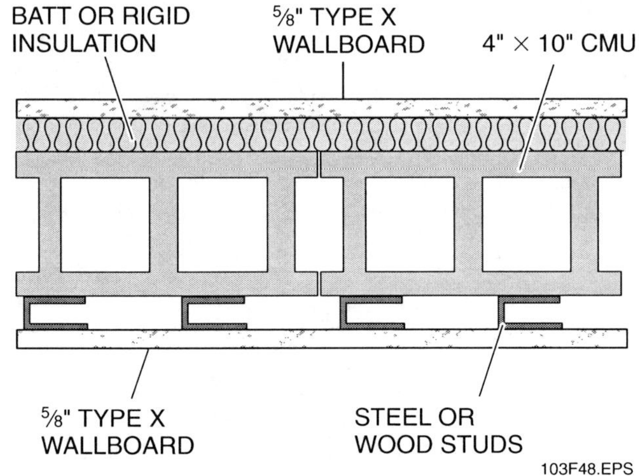

Figure 48   High fire/noise–resistant partition.

4.1.0 Fire-Rated Construction

Every wall, floor, and ceiling in a building is rated for its fire resistance, as established by building codes. The fire rating is stated in terms of hours, such as a one-hour wall or a two-hour wall. The rating indicates the length of time an assembly can withstand fire and provide protection from it, as determined under laboratory conditions (*Figure 49*). The higher the fire rating, the thicker the wall is likely to be.

In multi-family residential construction, such as apartments and townhouses, the walls and ceilings dividing the occupancies must meet special fire and soundproofing requirements. The code requirements vary from one location to another

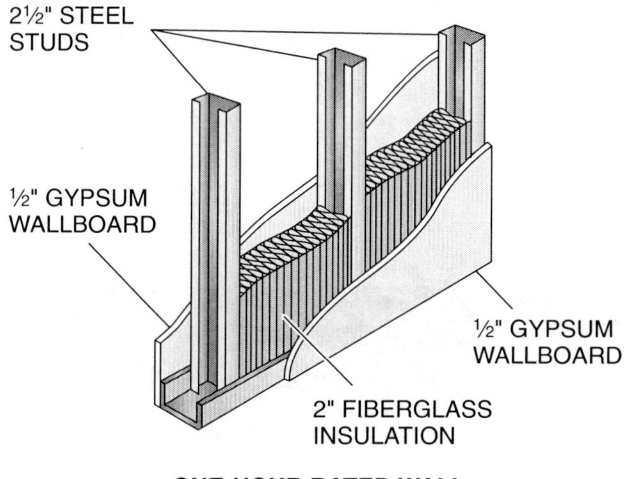

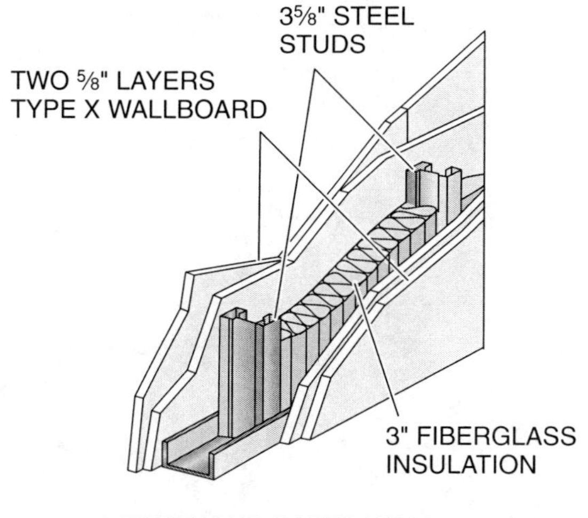

Figure 49 Partition wall examples.

Code Compliance

Local codes specify the fire ratings that must be observed and maintained in different occupancies and uses. The codes may also specify the nailing pattern to be used on drywall panels. Check the local codes before proceeding. Also, keep in mind that electrical and plumbing installations must be inspected in order for the building to receive a certificate of occupancy. Before covering these installations with drywall, make sure that the inspection has been performed. An inspection sheet should be available at the site.

and may even vary within areas of a jurisdiction. For example, buildings in high-risk areas may have stricter standards than those in other parts of the same city or county.

A fire-rated wall may abut a nonrated partition or wall. When this occurs, the rated wall must be carried through to maintain the fire rating (*Figure 50*).

4.1.1 Firestopping

Firestopping involves cutting off the air supply so that fire and smoke cannot readily move from one location to another.

In commercial construction and some residential applications, firestopping material is used to close wall penetrations, such as those created to run conduit, piping, cabling, and air conditioning ducts. If those openings are not sealed, fire can travel through them in its search for oxygen.

To meet the fire rating standards set by the building and fire codes, the openings must be sealed. The firestopping methods used for sealing are classified as mechanical and nonmechanical.

Mechanical firestops are devices, such as the pipe sleeve shown in *Figure 51*, that mechanically seal the opening.

Nonmechanical firestops are fire-resistant materials, such as caulks and putties, that are used to fill the space around the conduit, piping, or cable. You may have to install various nonmechanical firestopping materials when drilling through fire-rated walls and floors. Holes and gaps affect the fire rating of a floor or wall. Properly filling these openings with firestopping materials maintains the rating. Firestopping materials are applied around all types of piping, electrical conduit, ductwork, and electrical and communication cables

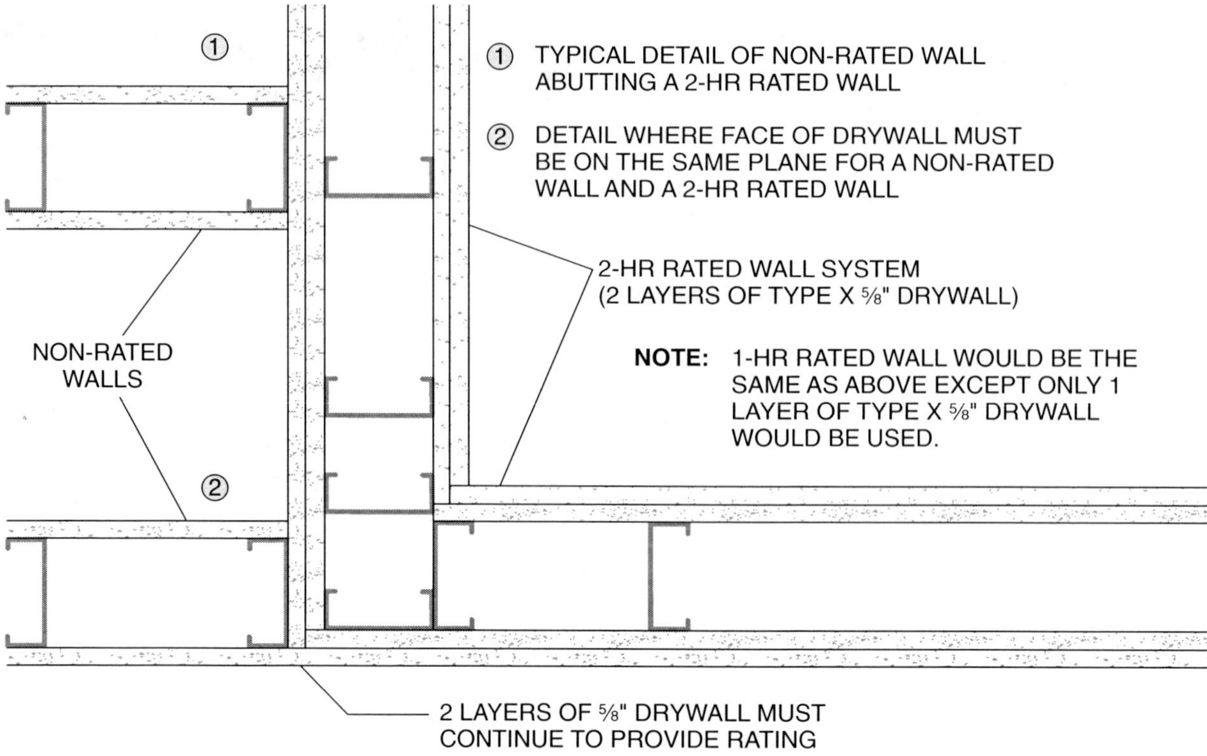

Figure 50   Example of a fire-rated wall abutting a nonrated wall.

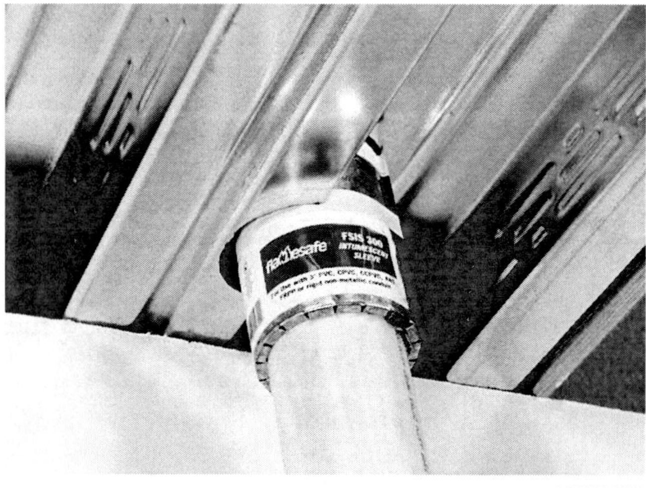

Figure 51   Firestop piping.

(*Figure 52*). They are also applied around similar devices that run through openings in floor slabs, walls, and other fire-rated building partitions and assemblies.

Nonmechanical firestopping materials are either intumescent or endothermic. Both types help control the spread of fire before, during, and after exposure to open flames. When subjected to the extreme heat of a fire, intumescent materials

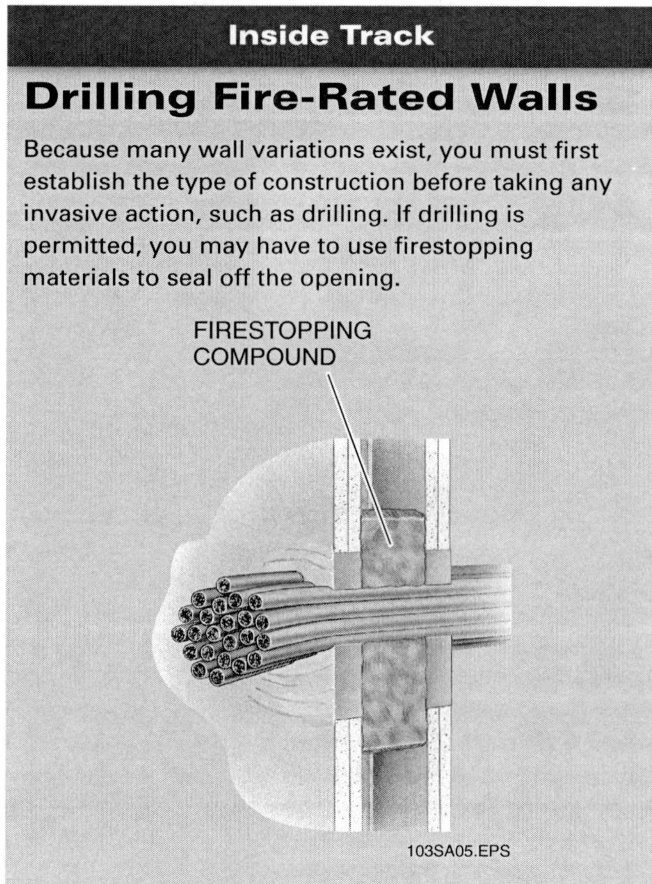

Drilling Fire-Rated Walls

Because many wall variations exist, you must first establish the type of construction before taking any invasive action, such as drilling. If drilling is permitted, you may have to use firestopping materials to seal off the opening.

FIRESTOP SEALANT

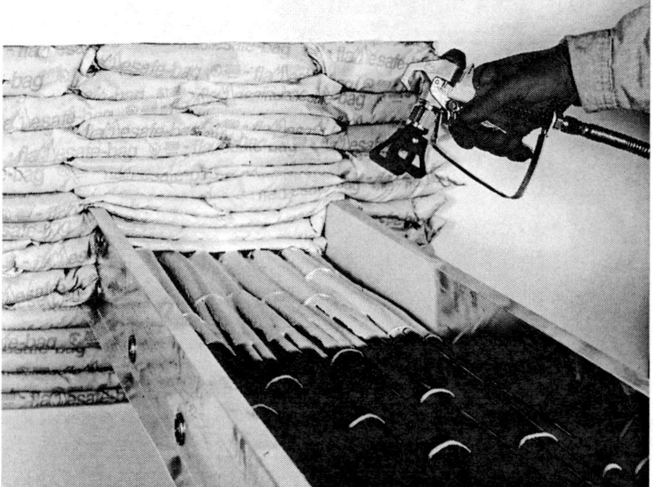

CABLE PROTECTION SPRAY

INTUMESCENT PUTTY

103F52.EPS

Figure 52 Fire barriers installed around electrical cables.

expand up to three times their original size to form a strong insulating material that seals the opening for three to four hours. If the insulation on the cables and pipes passing through the opening becomes consumed by the fire, the expansion of the firestopping material also acts to fill the void in the floor or wall. This expansion helps to stop the spread of smoke and other toxic products of combustion.

Endothermic materials block heat by releasing chemically bound water, which causes them to absorb heat.

Firestopping materials are free of corrosive gases when activated. This reduces the risks to building occupants and sensitive equipment.

Firestopping materials are available in a variety of forms. These forms include composite sheeting, caulks, silicone sealants, foams, moldable putty, wrap strips, and spray coatings. They come in both one-part and two-part formulas. The installation of these materials must always be done in accordance with the applicable building codes and the manufacturer's instructions for the product being used. Depending on the product, firestopping materials can be applied using spray equipment, caulking guns, pneumatic pumping equipment, or a putty knife.

Any firestopping materials used must meet the criteria of the American Society for Testing and Materials International (ASTM) standard *ASTM E814-08b, Standard Test Method for Fire Tests of Penetration Firestop Systems*, as tested under positive pressure. They must also have an hourly rating that is equal to or greater than the hourly rating of the floor or wall being penetrated. Based on *ASTM E814/UL 1479* tests, one of four ratings, measured in time, may be applied to firestopping materials and systems. These ratings are as follows:

- *F rating* – A firestopping system meets the requirements of an F rating if it remains in the opening during the fire test for the rating period without permitting the passage of flames through the opening or the occurrence of flaming on any element of the unexposed side of the assembly.
- *FT rating* – A firestopping system meets the requirements of an FT rating if it remains in the opening during the fire test within the limitations as specified for an F rating. In addition, the transmission of heat through the firestopping system during the rating period shall not have been such as to raise the temperature of any thermocouple on the unexposed surface of the firestopping system by more than 347.8°F (181°C) above its initial temperature.

- *FH rating* – A firestopping system meets the requirements of an FH rating if it remains in the opening during the fire test within the limitations for an F rating. In addition, during a hose-stream test, the firestopping system shall not develop any opening that would permit a projection of water from the stream to the unexposed side.
- *FTH rating* – A firestopping system shall be considered as meeting the requirements of an FTH rating if it remains in the opening during the fire test and hose stream test within the limitations as described for FT and FH ratings.

4.2.0 Sound-Isolation Construction

Buildings are generally required to meet a sound transmission class (STC) rating. The STC is a numeric rating that shows how effective the construction is in isolating airborne sound transmission. The higher the STC rating, the better the sound absorption. Hairline cracks and other openings can have an adverse effect on the ability of a building to achieve its STC rating, particularly in higher-rated construction. When a very high STC rating is needed, air conditioning, heating, and ventilating ducts should not be included in the assembly. Failure to observe special construction and design details can destroy the effectiveness of the best assembly. Techniques for improved sound isolation include the following:

- Separate framing for the two sides of a wall
- Resilient channel mounting for the gypsum board
- Using sound-absorbing materials in wall cavities
- Using adhesive-applied gypsum board of varying thicknesses in multi-layer construction
- Caulking the perimeter of gypsum board partitions, openings in walls and ceilings, partition/mullion intersections, and outlet box openings
- Locating recessed wall fixtures in different stud cavities

The entire perimeter of sound-isolating partitions is caulked around the gypsum board edges to make it airtight. The caulking should be a resilient sealant that does not harden, shrink, bleed, or stain.

Sound-control sealing must be covered in the specifications, understood by all related tradespersons, supervised by the appropriate party, and carefully inspected as the construction progresses.

5.0.0 FASTENERS AND ANCHORS

Conduit, cable clamps, raceways, and electronic assemblies must be secured to the structural components of a commercial building. Attaching something to concrete or steel usually involves first drilling a hole. Commercial structures require the use of different fasteners than those used in wood structures. Threaded screws and bolts are used in steel structures, for example. Concrete requires special anchors that are inserted into drilled holes and then expanded to secure them in place.

The two main kinds of fasteners are threaded and nonthreaded fasteners. There are many different types and sizes of fasteners within each of these two categories. Each type of fastener is designed for a specific use. The kind used for a given job may be listed in the project specifications, or you may be responsible for selecting an appropriate fastener.

Failure of fasteners to function properly can result in injury to people or damage to equipment. Performing quality work requires using the correct type and size of fastener for the job and knowing how to install it properly.

5.1.0 Threaded Fasteners

Threaded fasteners (*Figure 53*) are the most commonly used type of fastener. Many threaded fasteners are assembled with nuts and washers. The following sections describe the standard threads

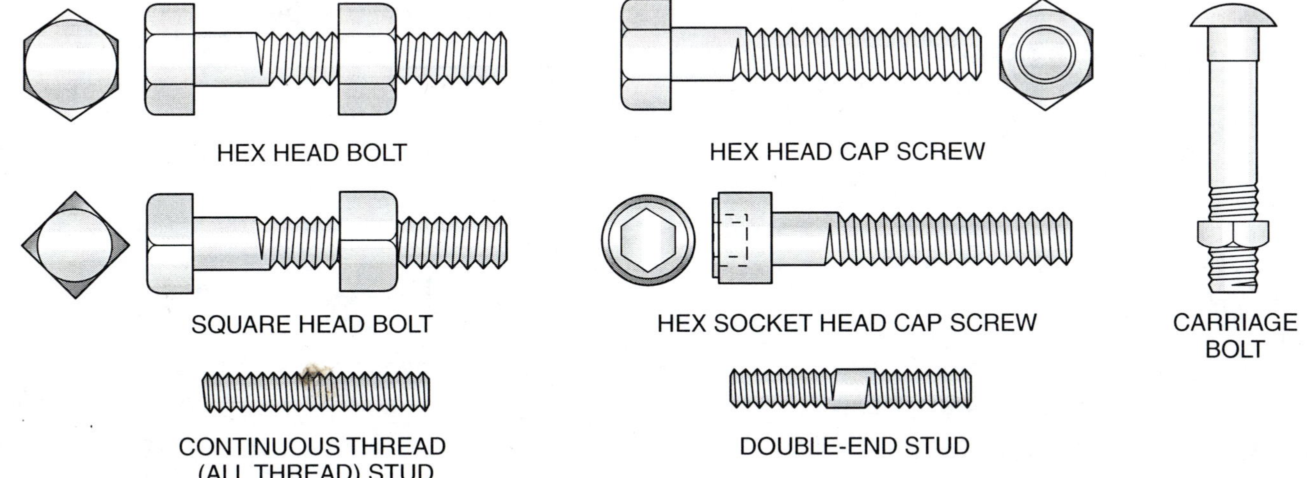

Figure 53 Threaded fasteners.

used on threaded fasteners, as well as the different types of bolts, screws, nuts, and washers used with the fasteners.

Many different types of threads are used in manufacturing fasteners. The various types of threads are designed for different jobs. Threads used on fasteners are manufactured to industry-established standards of uniformity. The most common thread standard is the Unified standard, sometimes referred to as the American standard. Unified standards are used to establish thread series and classes.

5.1.1 Thread Series

Unified standards exist for three series of threads, depending on the number of threads per inch for a specified diameter of fastener. The three thread series are as follows:

- *Unified National Coarse (UNC) thread* – Used for bolts, screws, nuts, and other general-purpose applications. Fasteners with UNC threads are commonly used for rapid assembly or disassembly of parts and in situations where corrosion or slight damage may occur.
- *Unified National Fine (UNF) thread* – Used for bolts, screws, nuts, and other applications in which a finer thread and a tighter fit are desired.
- *Unified National Extra Fine (UNEF) thread* – Used on thin-walled tubes, nuts, ferrules, and couplings.

5.1.2 Thread Classes

The Unified standards also establish thread classes. Classes 1A, 2A, and 3A apply only to external threads. Classes 1B, 2B, and 3B apply only to internal threads. Thread classes are distinguished from one another by the amount of

Thread Classes

External and internal thread classes differ widely with regard to tolerances. Why would you use Classes 3A or 3B for high-grade commercial products, such as precision tools and machines? What household products would warrant the use of Classes 1A and 1B?

tolerance provided. Classes 3A and 3B provide a minimum clearance, while classes 1A and 1B provide a maximum clearance.

Classes 2A and 2B are the most common. Classes 3A and 3B are used when close tolerances are needed. Classes 1A and 1B are used when fast and easy assembly is needed and a large tolerance is acceptable.

5.1.3 Thread Identification

Thread identification is done using a standard method. *Figure 54* shows screw thread designations for a common fastener. These thread designations are also covered in the following list:

- Nominal size – The approximate diameter of the fastener.
- *Number of threads per inch (TPI)* – The TPI is standard for all diameters.
- *Thread series symbol* – Indicates the Unified standard thread type (UNC, UNF, or UNEF).
- *Thread class symbol* – The closeness of fit between the bolt threads and nut threads.
- *Left-hand thread symbol* – Specified by the symbol LH. Unless threads are specified with the LH symbol, they are right-hand threads.

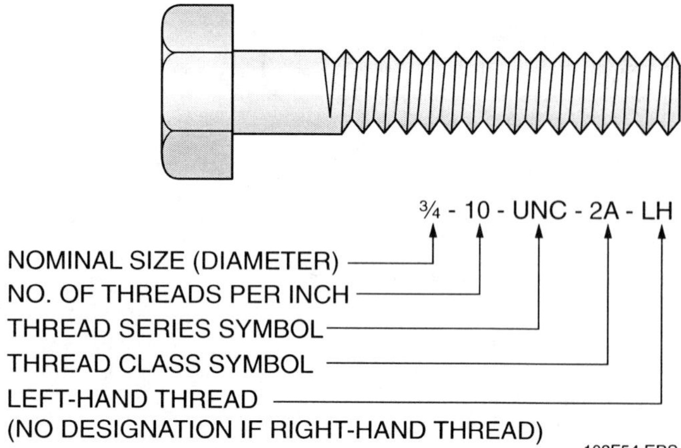

Figure 54 Screw thread designations.

Screw Diameters

A screw's diameter and TPI for a given project depend on what conditions?

5.1.4 *Grade Markings*

Special markings on the head of a bolt or screw can be used to determine the quality of the fastener. The Society of Automotive Engineers (SAE) and ASTM International have developed the standards for these markings. The grade or line markings for steel bolts and screws are shown in *Figure 55*.

Higher-quality steel fasteners have a larger number of marks on the head. If the head is unmarked, the fastener is usually considered to be made of mild steel (low carbon content).

5.2.0 Bolt and Screw Types

Bolts and screws are made in many different sizes and shapes and from a variety of materials. They are usually identified by the head type or other features. The following sections describe various types of bolts and screws.

5.2.1 *Machine Screws*

Machine screws (*Figure 56*) are used for general assembly work. They come in a number of different types and have slotted or recessed heads. Machine screws are available in diameters ranging from 0 (0.060") to ½" (0.500"). The length of machine screws ranges from ⅛" to 3", and they are also manufactured in metric sizes.

ASTM AND SAE GRADE MARKINGS FOR STEEL BOLTS & SCREWS

GRADE MARKING	SPECIFICATION	MATERIAL
	SAE-GRADE 0	STEEL
	SAE-GRADE 1 ASTM-A 307	LOW CARBON STEEL
	SAE-GRADE 2	LOW CARBON STEEL
	SAE-GRADE 3	MEDIUM CARBON STEEL, COLD WORKED
A 449	SAE-GRADE 5	MEDIUM CARBON STEEL, QUENCHED AND TEMPERED
	ASTM-A 449	
A 325	ASTM-A 325	MEDIUM CARBON STEEL, QUENCHED AND TEMPERED
BB	ASTM-A 354 GRADE BB	LOW ALLOY STEEL, QUENCHED AND TEMPERED
BC	ASTM-A 354 GRADE BC	LOW ALLOY STEEL, QUENCHED AND TEMPERED
	SAE-GRADE 7	MEDIUM CARBON ALLOY STEEL, QUENCHED AND TEMPERED ROLL THREADED AFTER HEAT TREATMENT
	SAE-GRADE 8	MEDIUM CARBON ALLOY STEEL, QUENCHED AND TEMPERED
	ASTM-A 354 GRADE BD	ALLOY STEEL, QUENCHED AND TEMPERED
A 490	ASTM-A 490	ALLOY STEEL, QUENCHED AND TEMPERED

ASTM SPECIFICATIONS
A 307 – LOW CARBON STEEL EXTERNALLY AND INTERNALLY THREADED STANDARD FASTENERS.
A 325 – HIGH STRENGTH STEEL BOLTS FOR STRUCTURAL STEEL JOINTS, INCLUDING SUITABLE NUTS AND PLAIN HARDENED WASHERS.
A 449 – QUENCHED AND TEMPERED STEEL BOLTS AND STUDS.
A 354 – QUENCHED AND TEMPERED ALLOY STEEL BOLTS AND STUDS WITH SUITABLE NUTS.
A 490 – HIGH STRENGTH ALLOY STEEL BOLTS FOR STRUCTURAL STEEL JOINTS, INCLUDING SUITABLE NUTS AND PLAIN HARDENED WASHERS.
SAE SPECIFICATION
J 429 – MECHANICAL AND QUALITY REQUIREMENTS FOR THREADED FASTENERS.

103F55.EPS

Figure 55 Grade markings for steel bolts and screws.

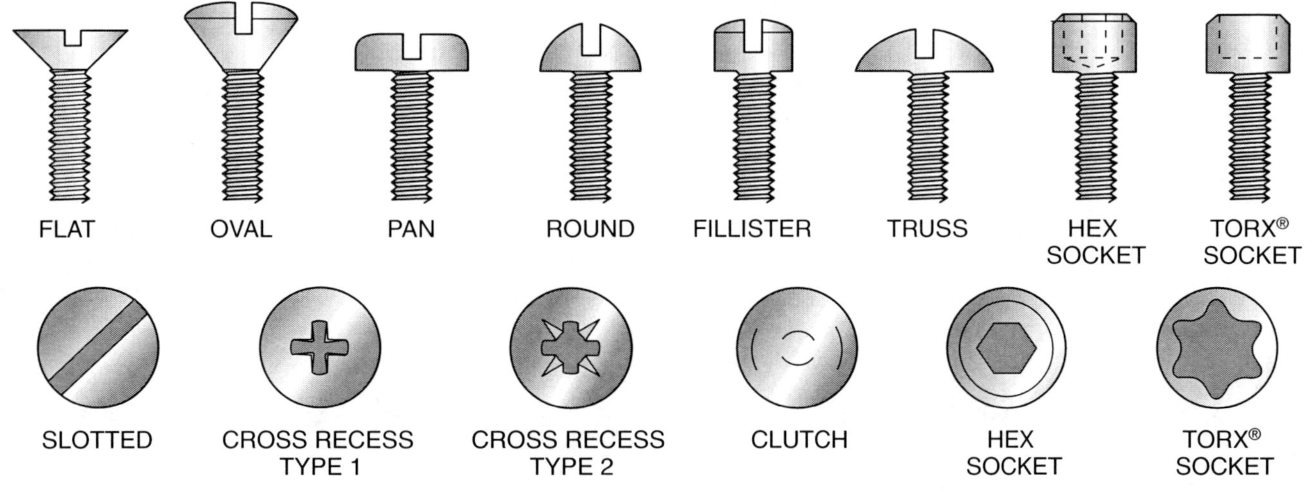

Figure 56 Machine screws.

The heads of machine screws are made in different shapes and have slots designed to fit various kinds of manual and powered screwdrivers. Flat-head screws are used in countersunk holes. They are tightened so that the head is flush with the surface. Oval-head screws are also used in countersunk holes where a more decorative finish is desired. Pan-head and round-head screws are general-purpose fastening screws. Fillister, hex socket, and TORX® socket screws are typically used on machined assemblies that require a finished appearance. They are often installed in a recessed hole. Truss screws are low-profile screws generally used without a washer. To prevent damage when tightening or removing machine screws (regardless of head type), always use a screwdriver or power tool bit with the proper tip for the job.

5.2.2 Machine Bolts

Machine bolts (*Figure 57*) are used to assemble parts in which close tolerances are not required. Machine bolts have square or hex heads and are available in diameters ranging from ¼" to 3". The length of machine bolts typically varies from ½" to 30". Nuts used with machine bolts are similar in shape to the bolt heads. The nuts are usually purchased at the same time as the bolts.

5.2.3 Cap Screws

Cap screws (*Figure 58*) are often used on high-quality assemblies requiring a finished appearance. The cap screw passes through a clearance hole in one of the assembly parts and screws into a threaded hole in the other part. The clamping action occurs by tightening the cap screw.

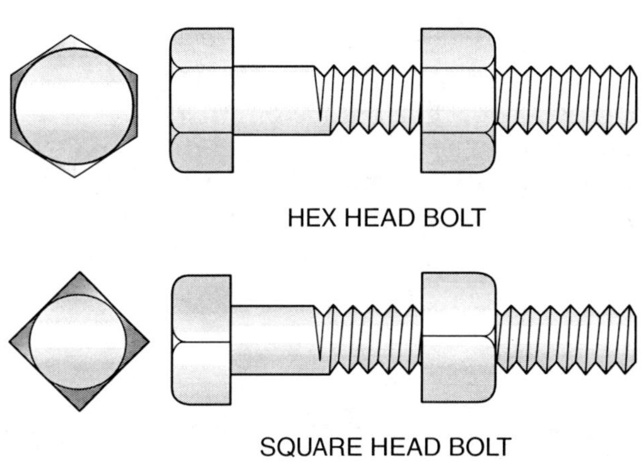

Figure 57 Machine bolts.

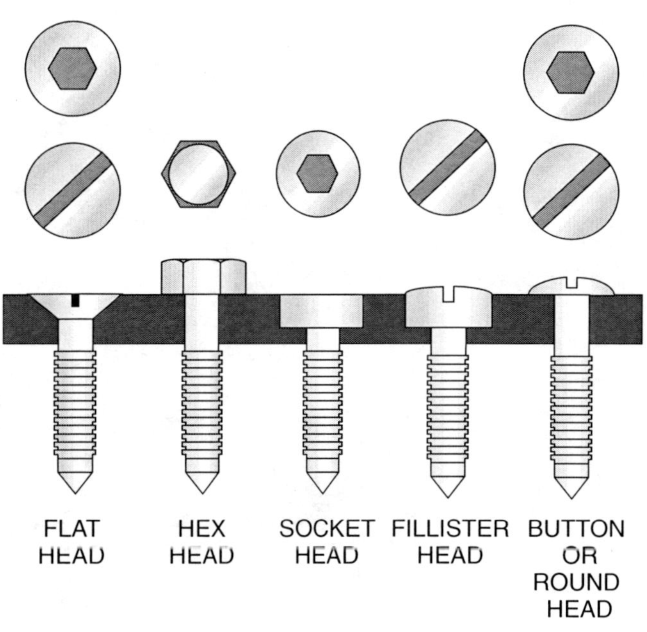

Figure 58 Cap screws.

Cap screws are made to close tolerances and have a machined or semi-finished bearing surface under the head. They are normally made in coarse and fine thread series and in diameters from ¼" to 2". Lengths may range from ⅜" to 10". Metric sizes are also available.

5.2.4 Setscrews

Heat-treated steel is normally used to make setscrews. In electrical work, setscrews are often used to secure EMT conduit fittings to the conduit sections. They are also used to prevent pulleys from slipping on shafts, hold collars in place on shafts, and hold shafts in place. The head style and point style are typically used to classify setscrews. *Figure 59* shows several setscrew head and point styles.

5.2.5 Stud Bolts

Stud bolts (*Figure 60*) are headless bolts that are threaded over the entire length of the bolt or threaded for a length on both ends of the bolt. One end of the stud bolt is screwed into a tapped hole. The part to be clamped is placed over the remaining portion of the stud, and a nut and washer are screwed on to clamp the two parts together. Other stud bolts have machine-screw threads on one end and lag-screw threads on the other so they can be screwed into wood.

5.3.0 Nuts

Most nuts used with threaded fasteners are hexagonal or square. Nuts are usually used with bolts that have the same head shape. *Figure 61* shows

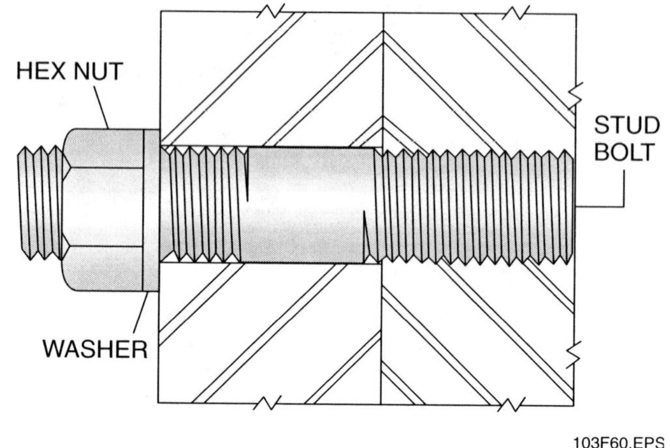

Figure 60   Stud bolt.

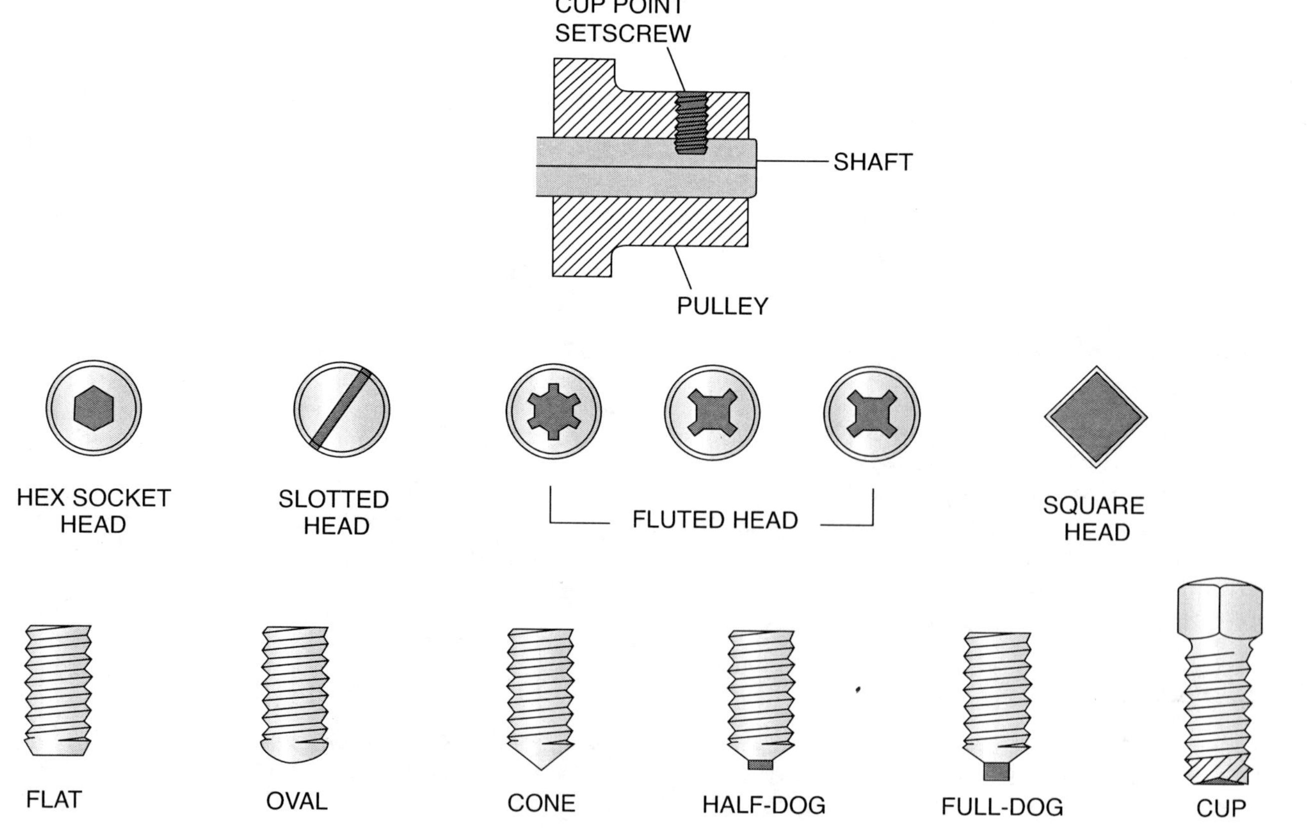

Figure 59   Setscrews.

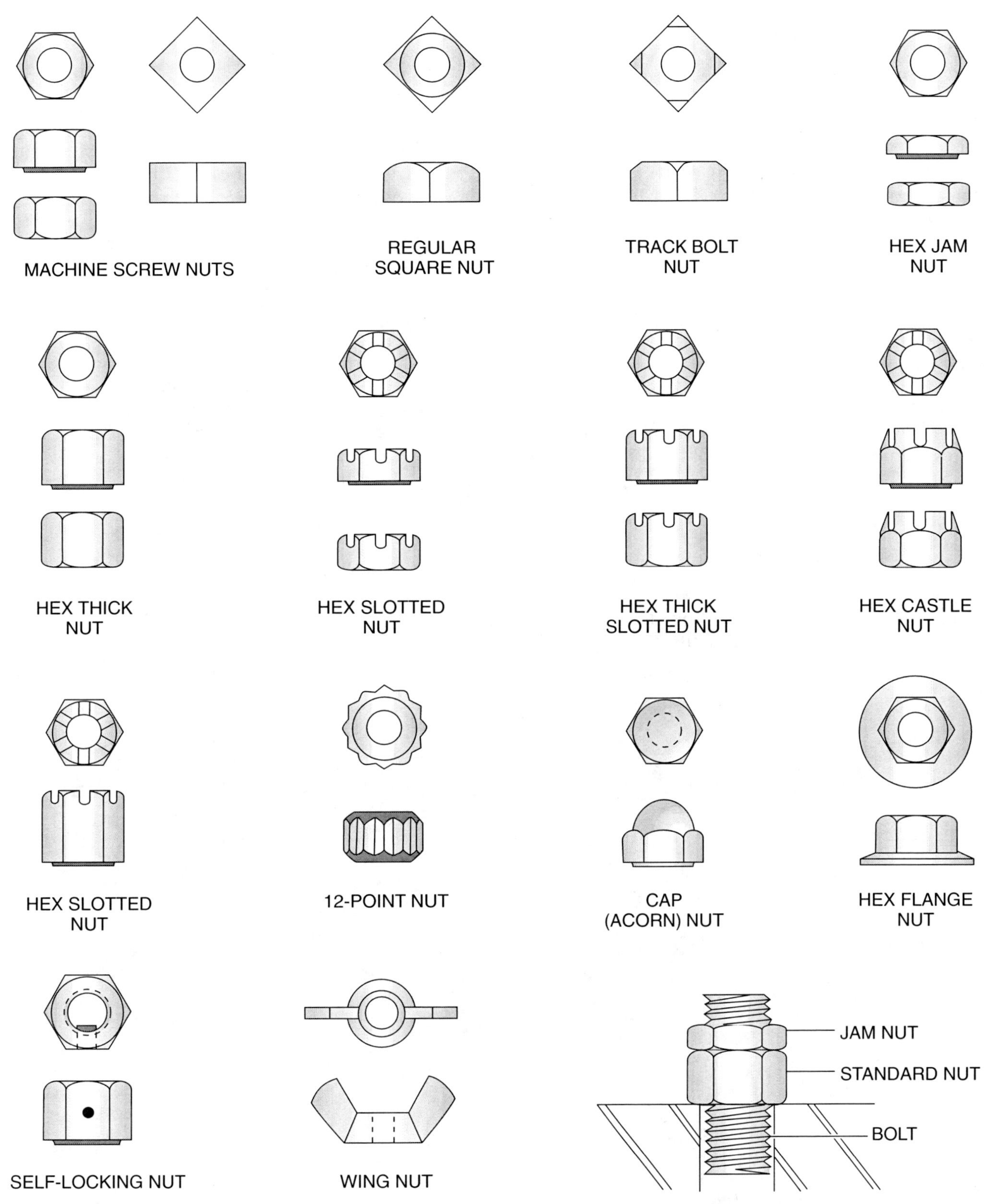

Figure 61   Nuts.

several different types of nuts used with threaded fasteners.

Nuts are classified as regular, semi-finished, or finished. The only machining done on regular nuts is to the threads. In addition to the threads, semi-finished nuts are machined on the bearing face. Machining the bearing face produces a truer surface for fitting the washer. The only difference between semi-finished and finished nuts is that finished nuts are made to closer tolerances.

The standard machine screw nut has a regular finish. Regular and semi-finished nuts are shown in *Figure 62*.

5.3.1 Jam Nuts

A jam nut is used to lock a standard nut in place. A jam nut is a thin nut installed on top of a standard nut. *Figure 63* shows a jam nut installation. Note that a regular nut can also be used as a jam nut.

5.3.2 Castellated, Slotted, and Self-Locking Nuts

Castellated (castle) and slotted nuts (*Figure 64*) are slotted across the flat part of the nut. They are used with specialty bolts where little or no loosen-ing of the fastener can be tolerated. After the nut has been tightened, a cotter pin is fitted in through a hole in the bolt and one set of slots in the nut. The cotter pin keeps the nut from loosening under working conditions.

Self-locking nuts (*Figure 64*) are also used where loosening of the fastener cannot be tolerated. Self-locking nuts are designed with nylon inserts, or they are purposely deformed in such a way that they cannot work loose. An advantage of self-locking nuts is that no hole is needed in the bolt.

5.3.3 Acorn Nuts

Acorn (cap) nuts (*Figure 65*) are used when appearance is important or when exposed, sharp thread edges on the fastener must be avoided. The acorn nut tightens onto the bolt and covers the ends of the threads. The tightening of an acorn nut is limited by the depth of the nut.

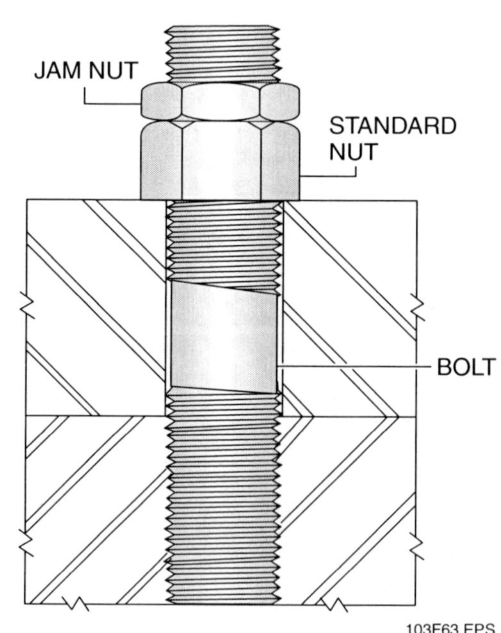

Figure 63   Jam nut.

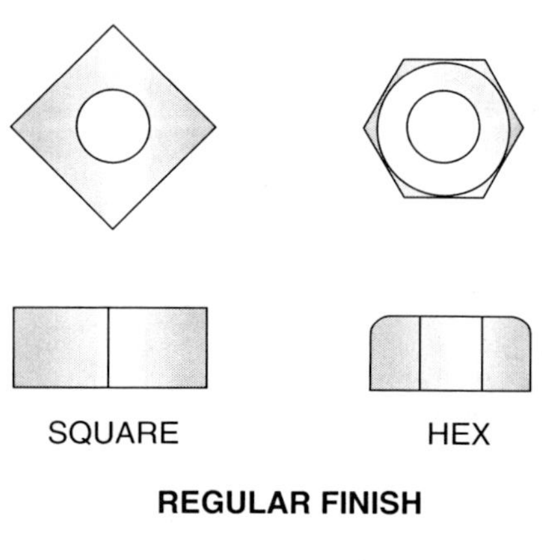

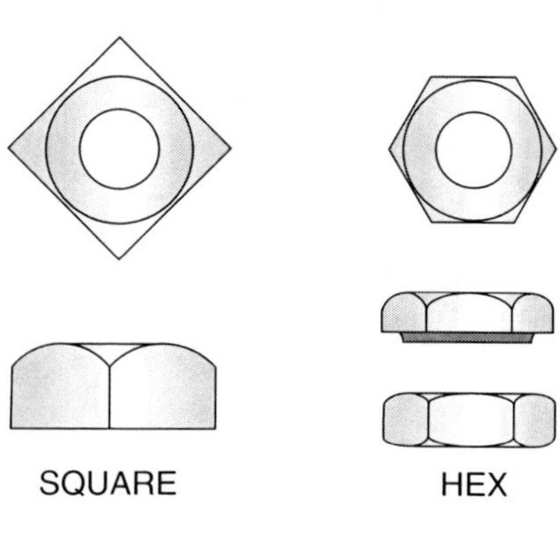

Figure 62   Nut finishes.

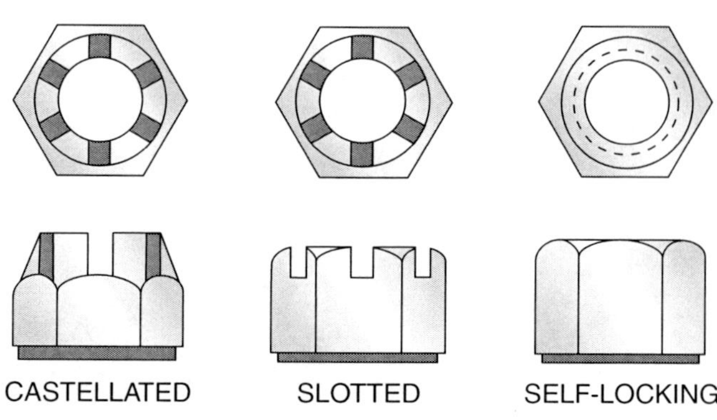

Figure 64   Castellated, slotted, and self-locking nuts.

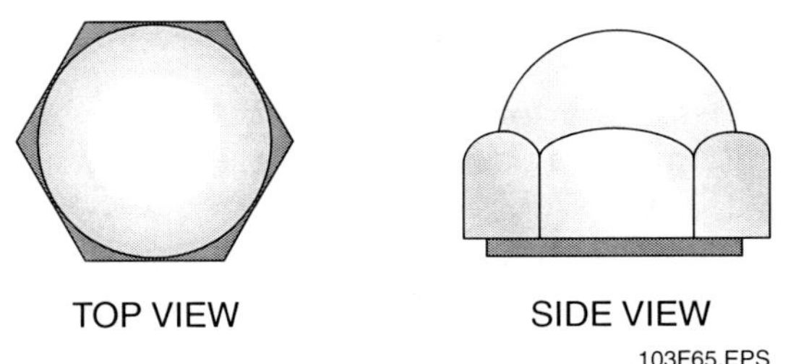

Figure 65   Acorn nut.

5.3.4 Wing Nuts

Wing nuts (*Figure 66*) are designed to allow rapid loosening and tightening of the fastener without the need for a wrench. Wing nuts are used where limited torque is required and frequent adjustments and service are necessary.

> NOTE
>
> Wing nuts should be used where hand tightening is sufficient.

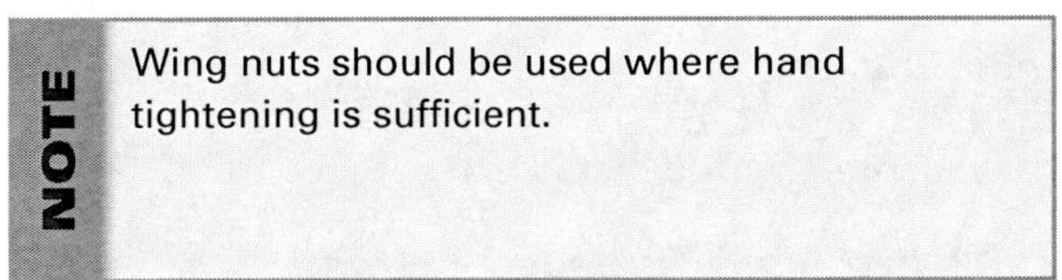

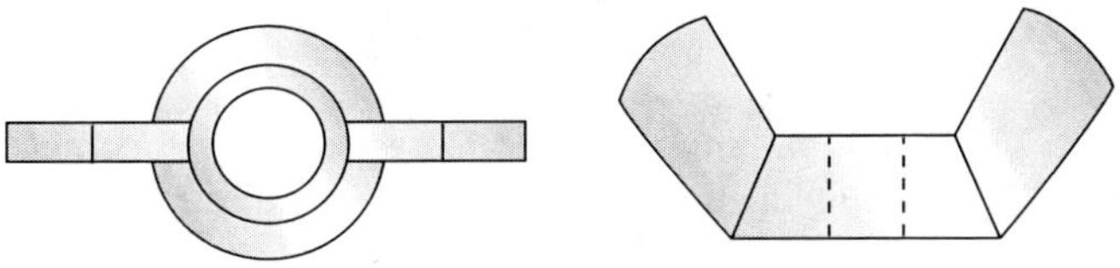

Figure 66   Wing nut.

5.4.0 Washers

Washers are available in several different types and sizes (*Figure 67*). Washers fit over a bolt or screw to provide an enlarged surface for bolt heads and nuts. They also serve to spread the fastener load over a larger area and to prevent marring of surfaces. Standard washers are made in light, medium, heavy-duty, and extra-heavy-duty series.

> NOTE
>
> The threads of the bolt or screw should have minimal clearance from the hole in the washer.

5.4.1 Lock Washers

Lock washers are designed to prevent bolts or nuts from working loose. Various types of lock washers are available for different applications:

- *Split-ring* – Commonly used with bolts and cap screws.
- *External (star)* – Used for the greatest resistance.
- *Internal* – Used with small screws.
- *Internal-external* – Used for oversized mounting holes.
- *Countersunk* – Used with flat or oval-head screws.

5.4.2 Flat and Fender Washers

Flat washers are used under bolts or nuts to spread the load over a larger area and protect the

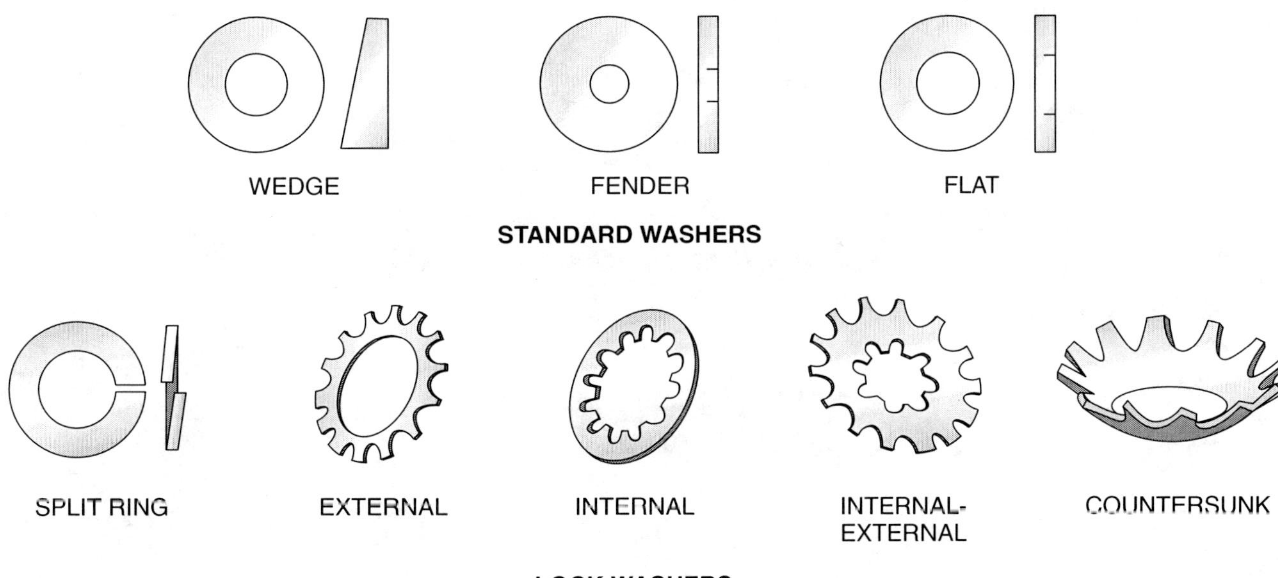

Figure 67   Washers.

surface. Common flat washers are made to fit bolt
or screw sizes ranging from No. 6 up to 1", with
outside diameters from ⅜" to 2".

Fender washers are wide-surface washers made
to bridge oversized holes or other wide clearances
in order to keep bolts or nuts from pulling through
the material being fastened. Fender washers are
flat and have a larger diameter and surface area
than regular washers. They may also be thinner
than a regular washer. Fender washers are typically made to fit bolt or screw sizes ranging
from 3/16" to ½" with outside diameters ranging from
¾" to 2".

5.5.0 Installing Fasteners

Different types of fasteners require different installation techniques. However, all installations
require knowing the proper installation methods,
tightening sequence, and torque specifications for
the type of fastener being used. Some bolts and
nuts require that special safety wires or pins be
installed to keep them from working loose.

Most fastener manufacturers provide charts
that specify the size hole that should be drilled
into the base material for each of their products
(*Figure 68*). The charts show the proper size drill
bit to use when it is necessary to first drill and
tap holes for use with machine bolts, screws, or
other threaded fasteners. The charts also show
the proper size drill bit to use for drilling pilot
holes used with metal and wood screws.

5.5.1 Torque Tightening

To properly tighten a threaded fastener, the two
factors that must be considered are the strength of
the fastener material and the degree to which the
fastener will be tightened.

A torque wrench (*Figure 69*) is used to control
the degree of tightness. The torque wrench mea-
sures how much a fastener is being tightened.
Torque is the turning force applied to the fastener.
Torque is normally expressed in inch-pounds (in-
lbs) or foot-pounds (ft-lbs). A one-pound force
applied to a wrench that is 1 foot long exerts
1 foot-pound, or 12 inch-pounds, of torque. The
torque reading is shown on an indicator on the
torque wrench as the fastener is being tightened.

Various types of bolts, nuts, and screws are
torqued to different values depending on the
application. Always check the project specifications and the manufacturer's manual to determine the proper torque for a particular type of
fastener. *Figure 70* shows selected torque values
for various graded steel bolts.

5.5.2 Installing Threaded Fasteners

The following general procedure can be used to
install threaded fasteners in a variety of applications:

NOTE

When installing threaded fasteners for a specific
job, always check all installation requirements.

WARNING!

To avoid injury, follow all safety precautions.

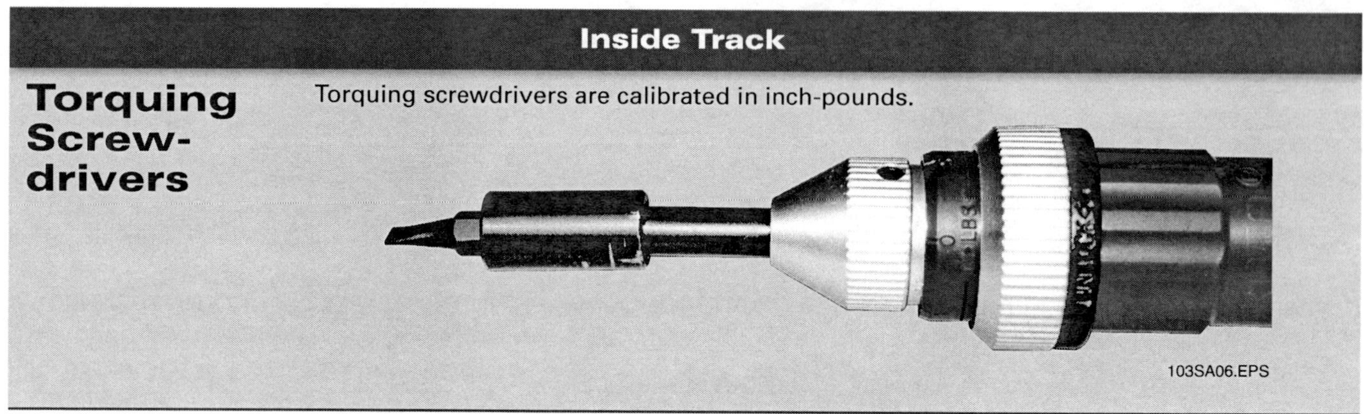

DRILL THIS SIZE HOLE		To Tap for This Size Bolt or Screw	For This Size Wood Screw Pilot in Hard Wood
Drill Size	Dec. Equiv.		
60	0.0400		
59	0.0410		
58	0.0420		
57	0.0430		
56	0.0465	0 × 80	
3/64	0.0469		
55	0.0520		
54	0.0550	1 × 56	No. 3
53	0.0595	1 × 64-72	
1/16	0.0625		
52	0.0635		No. 4
51	0.0670		
50	0.0700	2 × 56-64	
49	0.0730		No. 5
48	0.0760		
5/64	0.0781		
47	0.0785	3 × 48	No. 6
46	0.0810		
45	0.0820	3 × 56	
44	0.0860	4 × 36	No. 7
43	0.0890	4 × 40	
42	0.0935	4 × 48	
3/32	0.0937		
41	0.0960		
40	0.0980	5 × 36	No. 8
39	0.0995		
38	0.1015	5 × 40	
37	0.1040	5 × 44	No. 9
36	0.1069		
7/64	0.1094		
35	0.1100	6 × 32	
34	0.1110	6 × 36	
33	0.1130	6 × 40	No. 10
32	0.1160		
31	0.1200		No. 11
1/8	0.1250	7 × 36	
30	0.1285	8 × 30	No. 12
29	0.1360	8 × 32-36	
28	0.1405	8 × 40	
27	0.1440	9 × 30	
26	0.1470	3/16 × 24	
25	0.1495	10 × 24	No. 14
24	0.1520		
23	0.1540	10 × 28	
5/32	0.1562		
22	0.1570	10 × 30	
21	0.1590	10 × 32	
20	0.1610	3/16 × 32	
19	0.1660		
18	0.1695		No. 16
11/64	0.1719		
17	0.1730		
16	0.1770	12 × 24	
15	0.1800		
14	0.1820	12 × 28	
13	0.1850	12 × 32	No. 18
3/16	0.1875		
12	0.1890		

DRILL THIS SIZE HOLE		To Tap for This Size Bolt or Screw	For This Size Wood Screw Pilot in Hard Wood
Drill Size	Dec. Equiv.		
11	0.1910		
10	0.1935	15 × 20	
9	0.1960		
8	0.1990		
7	0.2010	1/4 × 20	
13/64	0.2031		
6	0.2040		
5	0.2055		
4	0.2090	1/4 × 24	No. 20
3	0.2130	1/4 × 28	
7/32	0.2187	1/4 × 32	
2	0.2210		
1	0.2280		No. 24
A	0.2340		
15/64	0.2344		
B	0.2380		
C	0.2420		
D	0.2460		
1/4	0.2500		

DRILL THIS SIZE HOLE		To Tap for This Size Bolt or Screw
Drill Size	Dec. Equiv.	
E	0.2500	
F	0.2570	5/16 × 18
G	0.2610	
17/64	0.2656	5/16 × 18
H	0.2660	
I	0.2720	
J	0.2770	5/16 × 24-32*
K	0.2810	
9/32	0.2812	5/16 × 24-32*
L	0.2900	
M	0.2950	
19/64	0.2969	
N	0.3020	
5/16	0.3125	3/8* × 16-1/8* P
O	0.3160	
P	0.3230	
21/64	0.3281	3/8 × 20-24
Q	0.3332	
R	0.3390	
11/32	0.3437	
S	0.3480	
T	0.3580	
23/64	0.3594	
U	0.3680	
3/8	0.3750	7/16 × 14
V	0.3770	
W	0.3860	
25/64	0.3906	7/16 × 14
X	0.3970	
Y	0.4040	
13/32	0.4062	
Z	0.4130	
27/64	0.4219	1/2 × 12-13
7/16	0.4375	1/4* Pipe
29/64	0.4531	1/2 × 20-24
15/32	0.4687	1/2 × 27
31/64	0.4844	9/16 × 12
1/2	0.5000	

* All tap drill sizes are for 75% full thread except asterisked sizes which are 60% full thread.

103F68.EPS

Figure 68 Fastener hole guide chart.

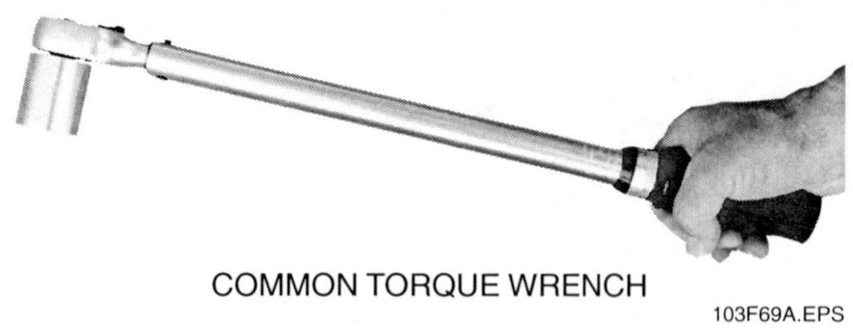

COMMON TORQUE WRENCH

103F69A.EPS

SATELLITE TV ANTENNA
TORQUE WRENCH

103F69B.EPS

Figure 69   Torque wrenches.

Step 1   Select the proper bolts or screws for the job.

Step 2   Check for damaged or dirty internal and external threads.

Step 3   Clean the bolt or screw threads. Do not lubricate the threads if a torque wrench is to be used to tighten the nuts.

Step 4   Insert the bolts through the predrilled holes and tighten the nuts by hand, or insert the screws through the holes and start the threads by hand.

> **NOTE**
>
> Turn the nuts or screws several turns by hand, and check for cross threading.

TORQUE IN FOOT POUNDS

FASTENER DIAMETER	THREADS PER INCH	MILD STEEL	STAINLESS STEEL 18-8	ALLOY STEEL
1/4	20	4	6	8
5/16	18	8	11	16
3/8	16	12	18	24
7/16	14	20	32	40
1/2	13	30	43	60
5/8	11	60	92	120
3/4	10	100	128	200
7/8	9	160	180	320
1	8	245	285	490

SUGGESTED TORQUE VALUES FOR GRADED STEEL BOLTS

GRADE		SAE 1 OR 2	SAE 5	SAE 6	SAE 8
TENSILE STRENGTH		64,000 PSI	105,000 PSI	130,000 PSI	150,000 PSI
GRADE MARK		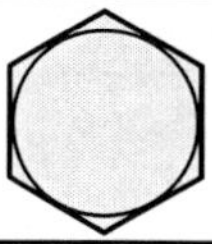	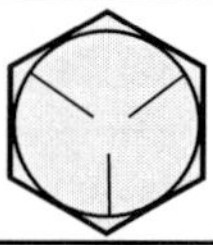	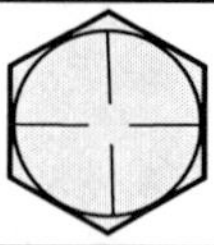	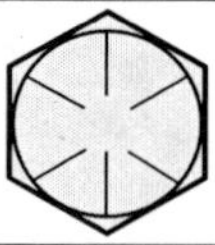
BOLT DIAMETER	THREADS PER INCH	FOOT POUNDS TORQUE			
1/4	20	5	7	10	10
5/16	18	9	14	19	22
3/8	16	15	25	34	37
7/16	14	24	40	55	60
1/2	13	37	60	85	92
9/16	12	53	88	120	132
5/8	11	74	120	169	180
3/4	10	120	200	280	296
7/8	9	190	302	440	473
1	8	282	466	660	714

103F70.EPS

Figure 70   Torque value chart.

Step 5 Follow the proper tightening sequence, and tighten the bolts or screws until they are snug.

Step 6 Check the torque specification. Following the proper tightening sequence, tighten each bolt, nut, or screw several times, approaching the specified torque. Tighten to the final torque specification.

Step 7 Install jam nuts, cotter pins, or safety wire if they are required to keep the bolts or nuts from working loose. *Figure 71* shows fasteners with a safety wire installed.

5.6.0 Eye Bolts

Eye bolts (*Figure 72*) get their name from the eye, or loop, at one end. The other end of an eye bolt is threaded. There are many types of eye bolts. The eye on some eye bolts is formed and welded, while the eye on other types is forged. Shoulder-forged eye bolts are often used as lifting devices and guides for wires, cables, and cords.

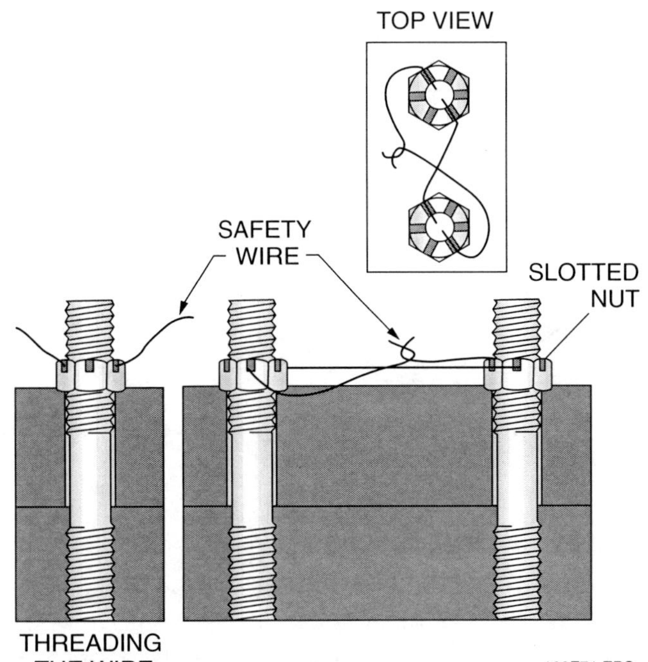

Figure 71 Safety-wired fasteners.

Inside Track

Taps and Dies

Taps and dies are used to cut threads in materials, such as metal, plastic, and hard rubber. Taps are used to cut internal threads in materials and dies are used to cut external threads on bolts and rods.

Taps and dies can be obtained in complete sets. The sets include various taps and dies, as well as diestocks, tap wrenches, guides, and the screwdrivers and wrenches necessary to loosen and tighten adjusting screws. This photo shows a tap and die set.

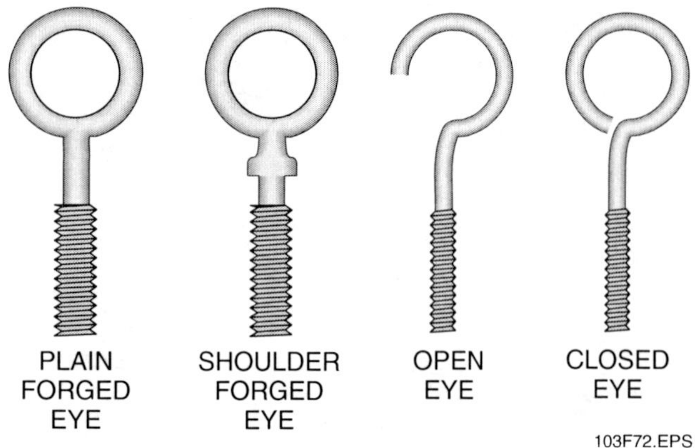

Figure 72 Eye bolts.

5.7.0 Hammer-Driven Pins and Studs

Hammer-driven pins or threaded studs (*Figure 73*) use a special tool to fasten wood or steel to concrete or block without the need to predrill holes. To install these fasteners, insert the pin or threaded stud into the hammer-driven tool point with the washer seated in the recess. Position the pin or stud against the base material to which it will be fastened, and tap the drive rod of the tool lightly until the striker pin contacts the pin or stud. Strike the tool's drive rod with about a two-pound engineer's hammer using heavy blows. The force of the hammer blows will transmit through the tool directly to the head of the fastener, causing it to be driven into the concrete or block. For best results, the drive pin or stud should be embedded to a minimum of ½" in hard concrete and to 1¼" in softer concrete block.

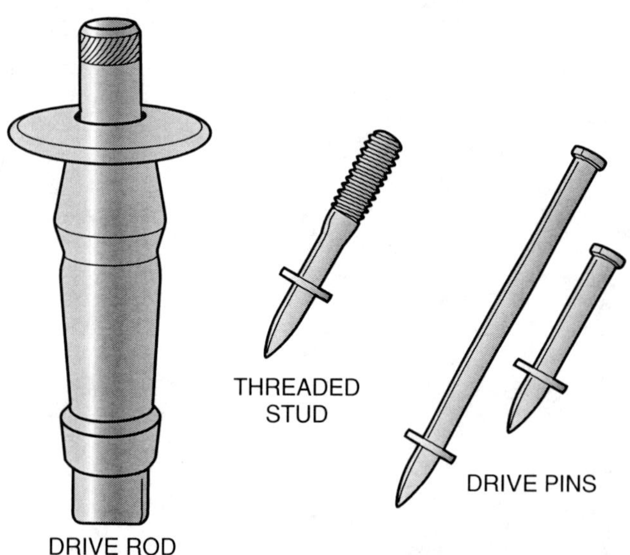

Figure 73 Hammer-driven pins and installation tool.

5.8.0 Mechanical Anchors

Mechanical anchors are used to give fasteners a firm grip in materials in which the fasteners by themselves would otherwise tend to pull out. Anchors are classified in many ways by different manufacturers. In this module, the categories used for anchors are as follows:

- One-step
- Bolt
- Screw
- Self-drilling

5.8.1 One-Step Anchors

One-step anchors can be installed through the mounting holes in the device to be fastened. This is because the anchor and the hole into which it is installed have the same diameter. One-step anchors come in diameters ranging from ¼" to 1¼", with lengths from 1¾" to 12". Common types of one-step anchors are wedge, stud, sleeve, one-piece, screw, nail, and Sammy® anchors (*Figure 74*).

- *Wedge anchors* – Heavy-duty anchors supplied with nuts and washers. The drill bit size used to drill the hole is the same diameter as the anchor. The depth of the hole is not critical as long as it meets the minimum recommended by the manufacturer. First, blow the hole clean of dust and other material. Insert the anchor

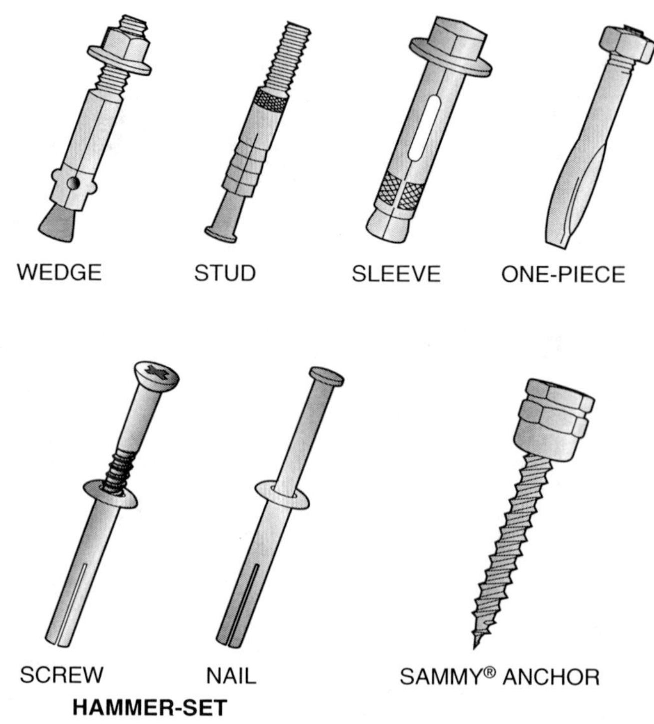

Figure 74 One-step anchors.

into the hole, and drive it with a hammer far enough so that at least six threads are below the top surface of the component. Then tighten the anchor nut to expand the anchor and secure it in the hole.

- *Stud bolt anchors* – Heavy-duty threaded anchors. Because this anchor is made to bottom in its mounting hole, it is a good choice when jacking or leveling of the fastened component is needed. The depth of the hole drilled in the masonry must be as specified by the manufacturer in order to achieve proper expansion. Blow the hole clean of dust and other material, and then insert the anchor in the hole with the expander plug end down. Drive the anchor into the hole with a hammer or setting tool to expand the anchor and tighten it in the hole. The anchor is fully set when it can no longer be driven into the hole. The component is fastened using a matching bolt.
- *Sleeve anchors* – Multi-purpose anchors in which the depth of the anchor hole is not critical as long as the minimum length recommended by the manufacturer is drilled. First, blow the hole clean of dust and other material. Insert the anchor into the hole, and tap it until it is flush with the component. Then tighten the anchor nut or screw to expand the anchor in the hole.
- *One-piece anchors* – Multi-purpose anchors that expand. As the anchor is driven into the hole, the spring force of the expansion mechanism is compressed and flexes to fit the size of the hole. Once set, it tries to regain its original shape. The hole drilled in the masonry must be at least one-half inch deeper than the required embedment. The proper depth is critical. Be sure not to overdrill or underdrill. First, blow the hole clean of dust and other material. Insert the anchor through the component, and drive it into the hole with a hammer until the head is firmly seated against the component. Make sure that the anchor is driven to the proper embedment depth. Manufacturers also make specially designed drivers and manual tools that can be used instead of a hammer to drive one-piece anchors. These tools allow the anchors to be installed in confined spaces. They also help prevent damage to the component from stray hammer blows.
- *Hammer-set anchors* – Made for use in concrete and masonry. There are two types: nail and screw. An advantage of the screw-type anchor is that it is removable. Both types have a diameter that is the same size as the anchoring hole. For both types, the anchor hole must be drilled to the diameter of the anchor and to a depth at least a one-quarter inch deeper than that required for embedment. First, blow the hole clean of dust and other material. Insert the anchor into the hole through the mounting holes in the component to be fastened. Then drive the screw or nail into the anchor body to expand it. Make sure that the head is seated firmly against the component and is at the proper embedment.
- *Sammy® anchors* – Are available for installation in concrete, steel, or wood. The Sammy® anchor is designed to support a threaded rod, which is screwed into the head of the anchor after it has been installed. A special nut driver is available for installing the screws.

5.8.2 Bolt Anchors

Bolt anchors are designed to be installed flush with the surface of the base material. They are used with threaded machine bolts or screws. Some types can be used with a threaded rod. Commonly used bolt anchors include drop-in and single- and double-expansion anchors (*Figure 75*).

Drop-in anchors are used as heavy-duty anchors. There are two types of drop-in anchors. The first type is used in solid concrete and masonry. It has an internally threaded expansion anchor with a preassembled internal expander plug. The anchor hole must be drilled to the diameter and depth specified by the manufacturer. First, blow the hole clean of dust and other material. Insert the anchor into the hole, and tap it until it is flush with the surface. Drive the setting tool (supplied with the anchor) into the anchor to expand it. Position the component in place, and fasten it by threading and tightening the correct size of machine bolt or screw into the anchor.

The second type, called hollow-set drop-in anchors, are made for use in hollow concrete and masonry base materials. Hollow-set drop-in anchors have a slotted, tapered expansion sleeve and a serrated expansion cone. They come in various lengths compatible with the outer wall thickness of most hollow base materials. They

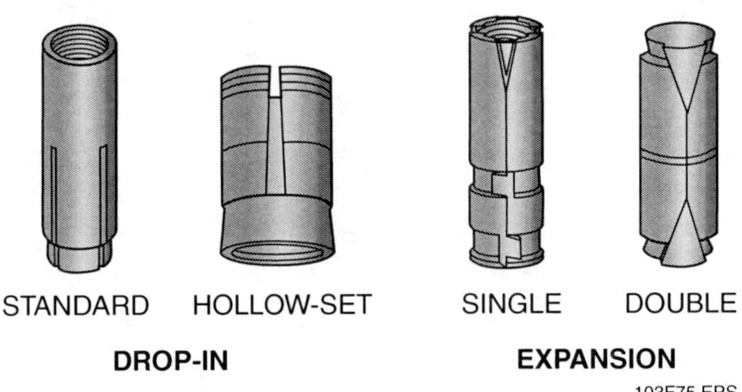

Figure 75 Bolt anchors.

can also be used in solid concrete and masonry. The anchor hole must be drilled to the diameter specified by the manufacturer. When installing in hollow base materials, drill the hole into the cell or void. Blow the hole clean of dust and other material, insert the anchor into the hole, and tap it until it is flush with the surface. Position the component to be fastened in place, and then thread the proper size of machine bolt or screw into the anchor. Tighten it to expand the anchor in the hole.

Single- and double-expansion anchors are made for use in concrete and other masonry. The double-expansion anchor is mainly used when fastening into concrete or masonry of questionable strength. For both types, the anchor hole must be drilled to the diameter and depth specified by the manufacturer. Blow the hole clean of dust and other material, and then insert the anchor into the hole, threaded cone end first. Tap the anchor until it is flush with the surface. Position the component to be fastened in place, and then thread the proper size machine bolt or screw into the anchor and tighten it to expand the anchor in the hole.

5.9.0 Self-Drilling Anchors

Some anchors used in masonry are self-drilling. The fastener shown in *Figure 76* has a cutting sleeve that is first used as a drill bit and later becomes the expandable fastener. The rotary hammer drills the hole in the concrete using the anchor sleeve as the drill bit.

After the hole has been drilled, the anchor is pulled out and the hole cleaned. Then the anchor's expander plug is inserted into the cutting end of the sleeve. The anchor sleeve and expander plug are driven back into the hole with the rotary hammer until they are flush with the surface of the concrete. As the fastener is hammered down, it hits bottom, and the tapered expander causes the fastener to expand and lock into the hole. The anchor is then snapped off at the shear point with a quick lateral movement of the hammer. The component to be fastened can then be attached to the anchor using the proper size of bolt.

5.10.0 Guidelines for Drilling Anchor Holes in Hardened Concrete and Masonry

When selecting masonry anchors, always follow the manufacturer's recommendations for the hole diameter and depth, minimum embedment in concrete, maximum thickness of the material to be fastened, and the pullout and shear load capacities.

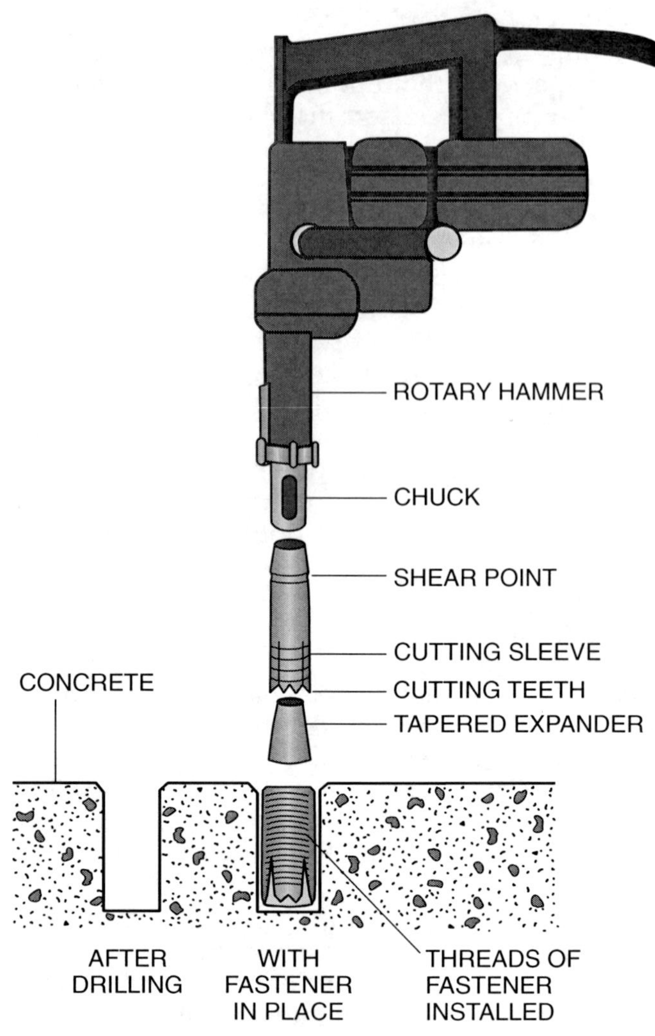

Figure 76   Self-drilling anchor.

When installing anchors or anchor bolts in hardened concrete, make sure that the area where the equipment or component will be fastened is smooth and provides a solid footing. Uneven footing could cause the equipment to twist, warp, not tighten properly, or vibrate when in operation. Before starting, carefully inspect the rotary hammer or hammer drill and the drill bits to ensure that they are working properly. Set the drill or hammer tool depth gauge to the required depth of the hole. Be sure to use the type of carbide-tipped masonry or percussion drill bits recommended by the manufacturer, because these bits are made to take the higher impact of the masonry materials. To use masonry drill bits properly, avoid forcing them into the material by pushing down too hard on the drill. Use a little pressure, and let the drill do the work. For large holes, start with a smaller bit, and then change to a larger bit.

The methods for installing the different types of anchors in hardened concrete or masonry were briefly described in previous sections of this module. Always install the selected anchors according to the manufacturer's directions. An example showing how many types of expansion anchors are installed in hardened concrete or masonry is shown in *Figure 77*.

WARNING!

Drilling into concrete generates noise, dust, and flying particles. Always wear safety goggles, ear protectors, and gloves. Make sure that other workers in the area also wear protective equipment.

Step 1 Drill the anchor bolt hole the same size as the anchor bolt. The hole must be deep enough so that six threads of the bolt are below the surface of the concrete. Clean out the hole using a squeeze bulb.

Step 2 Drive the anchor bolt into the hole using a hammer. Protect the threads of the bolt with a nut that does not allow any threads to be exposed.

Step 3 Put a washer and nut on the bolt, and tighten the nut with a wrench until the anchor is secure in the concrete.

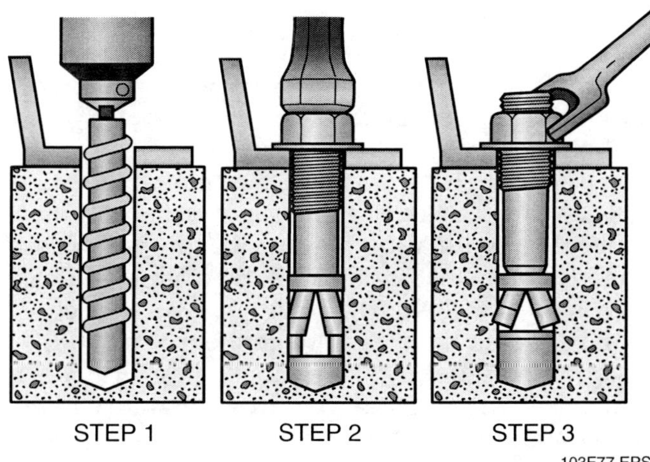

Figure 77 Installing an anchor bolt in hardened concrete.

5.11.0 Epoxy Anchoring Systems

Epoxy resin compounds can be used to anchor threaded rods, dowels, and similar fasteners in solid concrete, hollow wall, and brick. In one manufacturer's product, a two-part epoxy is packaged in a two-chamber cartridge that keeps the resin and hardener ingredients apart until use. This cartridge is placed into a special tool similar to a caulking gun. When the gun handle is pumped, the epoxy resin and hardener are mixed within the gun; then the epoxy is ejected from the gun nozzle.

To use epoxy to install an anchor in solid concrete (*Figure 78*), drill a hole of the proper size in the concrete and clean it using a nylon (not metal) brush. Dispense a small amount of epoxy from the gun to check that the resin and hardener have mixed properly. This is indicated by the epoxy being a uniform color. Place the gun nozzle into the hole and inject the epoxy until half the depth of the hole is filled. Push the selected fastener into the hole with a slow twisting motion to make sure that the epoxy fills all voids and crevices. Then set the fastener to the required plumb (or level) position. After the recommended curing time for the epoxy has elapsed, tighten the fastener nut to secure the component or fixture in place.

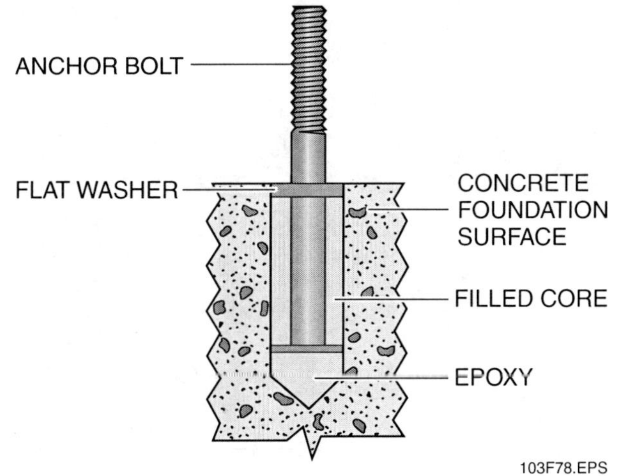

Figure 78 Fastener anchored in epoxy.

Use the Proper Tool for the Application

To avoid damaging fasteners, use the correct tool for the job. For example, avoid using pliers to install bolts or using screwdrivers that are too large or too small.

The procedure for installing a fastener in a hollow wall or brick when using epoxy is basically the same as the one just described. The difference is that the epoxy is first injected into an anchor screen to fill the screen, and then the anchor screen is installed into the hole. The anchor screen is needed to hold the epoxy intact in the hole until the anchor is inserted.

6.0.0 SPECIAL TOOLS

Special tools are needed to attach electronic devices, cable clamps, raceways, and conduit to a concrete or steel structure. The basic drills, power screwdrivers, and saws used in residential construction may not be suitable for most commercial construction.

6.1.0 Hammer Drills and Rotary Hammers

Hammer drills and rotary hammers (*Figure 79*) are special types of power drills. When equipped with a carbide-tipped percussion bit, they are used mainly to drill holes in concrete and other masonry materials. When equipped with different bits or cutters, they can also be used for the following purposes:

- Drilling holes in concrete, steel, and other materials
- Setting anchors
- Performing light chipping work

Hammer Drill Safety

Make sure that you have a firm grip on the side handle when using a hammer drill. Never hold on to just the main handle. Use both hands to equalize the rotation of the drill. Most hammer drills have enough torque to break your wrist.

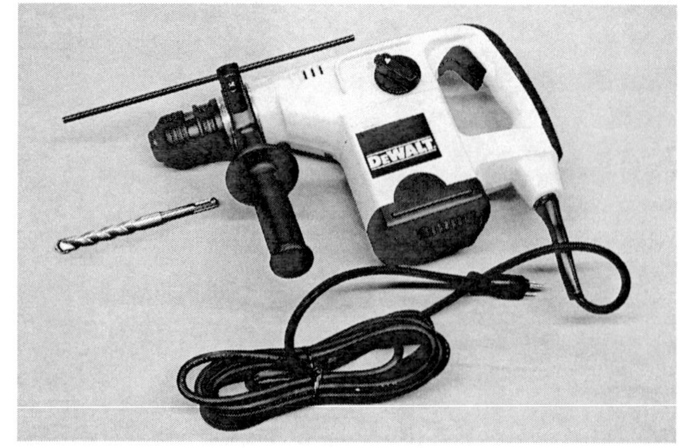

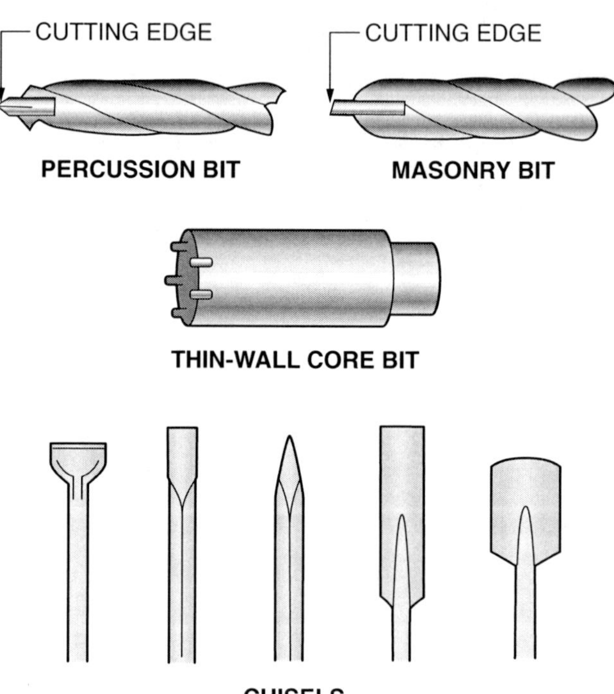

Figure 79 Rotary hammer and accessories.

The hammer drill is a lighter-duty tool used for drilling smaller holes. The rotary hammer is a heavier-duty tool used to drill larger holes. Both the hammer drill and rotary hammer operate with a dual action. They rotate and hammer at the same time, enabling them to drill holes much faster than can be done with a standard drill equipped with a masonry bit. Most models are reversible and can be easily switched to a standard rotary-action drill. Both also have a depth gauge that can be set to control the depth of the hole being drilled. A cold chisel bit can be used with some rotary hammers for chipping and edging concrete. Chipping hammers, similar to rotary hammers, are made for use in chipping masonry and concrete.

A heavy-duty rotary hammer can accommodate a 6" core bit, as well as a variety of chisels used to penetrate concrete and stone.

As always, safety is the primary concern when using power tools. Guidelines for the safe use of portable drills and hammer drills are as follows:

- Hold the tool firmly.
- Always remove the chuck key before starting the drill.
- Be sure the drill or tool is secure in the chuck before starting the drill.
- Use appropriate eye protection.
- Never attempt to stop the drill by taking hold of the chuck.
- Avoid forcing the drill into the material.
- Be sure that the material is properly secured before drilling.
- Never point the drill at anyone.

Because of the wide variety of applications and tools used to drill or cut through various types of materials, two tables are provided for reference. *Table 2* cross-references various types of materials to the tools that can be used to bore or cut through them. *Table 3* relates different types and sizes of bits and blades to the tools with which they can be used.

6.2.0 Core Drills

Core drills (*Figure 80*) are heavy-duty drills designed for drilling through reinforced concrete, brick, block, and stone. They are usually attached to the wall for horizontal drilling or mounted on a fixture for vertical drilling. Core drills with diamond bits can drill a 12"-diameter hole. The bit, which is similar to a hole saw, is cooled and lubricated by running water during drilling. This type of tool requires special training.

6.3.0 Metal Stud Punches

Metal studs usually come from the factory with prepunched openings for conduit and piping. A metal stud punch (*Figure 81*) is available for punching additional openings if needed. This tool works with studs up to 20 gauge. When electrical or telecommunications cabling is routed through the openings, grommets like the one shown should be inserted into the openings to protect the cabling from sharp edges. Other types of devices are made to eliminate conduit rattle.

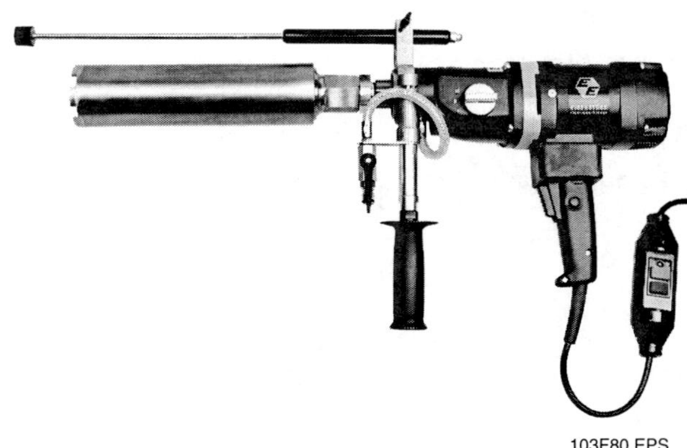

103F80.EPS

Figure 80　Core drill.

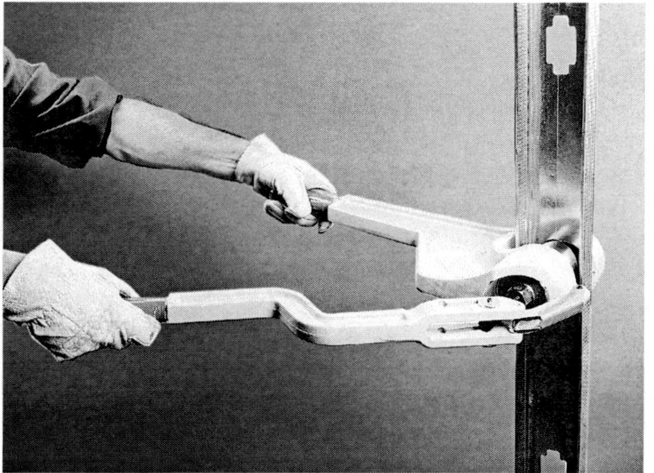

103F81.EPS

Figure 81　Metal stud punch and grommet.

Table 2 Drill Bits and Blades Used for Various Materials

TYPE & TYPICAL SIZES	SUGGESTED TOOL FOR CUT OR HOLE							
	Standard 3/8" & 1/2" Drill	Cordless 3/8" & 1/2" Drill	Hammer Drill	Close Quarter Drill	1/2" Hammer Drill	1/2" Hole Hawg Drill	Hand Auger	1/2" 90° Angle Drill
Twist Bits, Standard Shank from 1/16" – 1/2"	●	●	●	●	●			
Twist Bits, 1/2" – 1"	●	●	●		●			
Brad Point Wood Bit (same as twist bit size)	●	●	●	●	●			
Stepper Bits, 1/8" – 1 3/8"	●	●	●	●	●			
Spade Bits, 1/4" – 1 1/2"	●	●	●	●	●			
Easy Bore Bits, 1/2" – 4 5/8"	1/2" recommended	●	●		●			●
Bell Hanger Bits, 1/4" – 3/8"	●	●	●	●	●		●	
Hole Saws, 9/16" – 6"	1/2" required	●	●		●	●		●
Bimetal Hole Saws, 9/16" – 6"	1/2" required	●	●		●			●
Forstner Bits (Precision Boring), 1/4" – 2 1/8"	●	●	●		●			●
Carbide Easy Bore Bits, 3/4" – 2 1/2"						●		●
Auger Type, 9/16" – 1 1/2"	●				●	●		●
Around the Corner Bits (Cuts Curved Holes), 3/4" – 1 1/8"	●	●			●	●		●
Type "C" Combination Flex Bit, 4' – 6' Long 1/4" – 1" Dia.	1/2" HD					●		●
Type "B" Combination Flex Bit, 4' – 6' Long 3/8" – 9/16"	1/2" HD	●			●	●		●
Auger Flex Bit, 4' – 6' Long 3/8" – 1" Dia.	1/2" HD				●	●		●

MASONRY					
	Standard 3/8" & 1/2" Drill	Cordless Drill	Hammer Drill	Rotary Hammer	Demolition Hammer
Masonry Twist Bit from 3/16" to 1/2"	●	●	●		
Percussion Carbide Bit from 3/16" to 1/2"			●		
Type "M" Flex Bits from 1/4" to 3/4"			●		
Hammer Core Bits from 2" to 6"				●	●

103T02.EPS

Table 3 Drill Bits and Blades Used with Various Tools

TYPE & TYPICAL SIZES	MATERIAL										
	Wood	Steel	Metal Studs	Vinyl Siding	Alum. Siding	Drywall	Plaster	Ceiling Tile	Formica & Laminates	Asphalt Roofing	Metal Siding
Twist Bits, Standard Shank from 1/16" – 1/2"	●	●	●	●	●	●		●	●	●	●
Twist Bits, 1/2" – 1"	●		●	●	●	●		●	●	●	●
Brad Point Wood Bit (same as twist bit size)	●			●		●		●	●	●	●
Knockout Punch, Various Sizes & Shapes		●	●								
Metal Stud Punch, 7/8" – 1⅜"			●								
Stepper Bits, 1/8" – 1⅜"		●	●		●						●
Spade Bits, 1/4" – 1½"	●					●		●	●		
Easy Bore Bits, 1/2" – 4⅝"	●					●				●	
Bell Hanger Bits, 1/4" – 3/8"	●			●	●	●		●	●	●	●
Hole Saws, 9/16" – 6"	●			●		●		●	●	●	
Bimetal Hole Saws, 9/16" – 6"	●	●	●	●	●	●		●	●	●	●
Forstner Bits (Precision Boring), 1/4" – 2⅛"	●			●		●		●	●		
Carbide Easy Bore Bits, 3/4" – 2½"	●			●	●	●	●	●		●	
Auger Type, 9/16" – 1½"	●					●			●		
Around the Corner Bits (Cuts Curved Holes), 3/4" – 1⅛"	●			●		●		●			
Type "C" Combination Flex Bit, 4' – 6' Long 1/4" – 1" Dia.	●		●	●		●		●		●	
Type "B" Combination Flex Bit, 4' – 6' Long 3/8" – 9/16"	●					●					
Auger Flex Bit, 4' – 6' Long 3/8" – 1" Dia.	●										
Wallboard Saw						●		●			
Jab Saw with Saws-All Blade	●			●		●	●	●			
Utility Knife						●					
Saws-All Blades, 3" – 12"	●	●	●	●	●	●	●	●	●	●	●

TYPE & TYPICAL SIZES	MASONRY						
	Concrete	Brick	Stone	Ceramic & Mosaic Floor & Wall Tile	Plaster with or without Wire Mesh	Corian & Granite	Block
Masonry Twist Bit from 3/16" to 1/2"	●	●		●	●	●	●
Percussion Carbide Bit from 3/16" to 1/2"	●	●	●	●	●	●	●
Type "M" Flex Bits from 1/4" to 3/4"	●			●	●		●
Hammer Core Bits from 2" to 6"	●	●	●				●

103T03.EPS

6.4.0 Powder-Actuated Tools and Fasteners

Powder-actuated tools (*Figure 82*) can be used to drive a wide variety of specially designed pin and threaded stud-type fasteners into masonry and steel. These tools look and fire like a gun and use the force of a gunpowder load (typically .22, .25, or .27 caliber) to drive the fastener into the material. The depth to which the pin or stud is driven is controlled by the density of the base material and the power level or strength of the powder load.

Powder loads and their cases are used with specific types and/or models of powder-actuated tools and are not interchangeable. Typically, powder loads are made in 12 increasing power or load levels in order to achieve the proper penetration. The different power levels are identified by a color-code system. Note that different manufacturers may use different color codes to identify load strength. Power level 1 is the lowest power level, while 12 is the highest. Power levels that are higher in number are used for driving into hard materials or when a deeper penetration is needed. Powder loads are available as single-shot units, for use with single-shot tools, and as multi-shot strips or disks for semiautomatic tools.

> **WARNING!**
>
> Powder-actuated fastening tools are only to be used by trained and licensed operators who must use them in accordance with the tool operator's manual. All operators must carry their licenses with them whenever they are using powder-actuated tools.

OSHA Standard 29 CFR 1926.302(e) governs the use of powder-actuated tools. It states that only those workers who have been trained to operate a particular powder-actuated tool are allowed to operate it. Authorized instructors available from the various powder-actuated tool manufacturers generally provide the training and licensing. When using powder-actuated driver tools, trained operators must take all appropriate safety precautions to protect both themselves and others in the area as follows:

- Always use the tool in accordance with the published tool operation instructions, which should be kept with the tool. Never attempt to override the safety features of the tool.
- Never place your hand or any other body parts over the front muzzle end of the tool.
- Use only fasteners, powder loads, and tool parts specifically made for use with the tool. Using other materials can cause improper and unsafe functioning of the tool.
- Operators and bystanders must wear eye and hearing protection and hard hats. Any other required personal safety gear must also be used.
- Always post warning signs that state, *Powder-Actuated Tool in Use*, within 50 feet of the area where the tools are being used.
- Prior to using a tool, make sure that it is unloaded and perform a proper function test. Check the functioning of the unloaded tool as described in the published tool operation instructions.

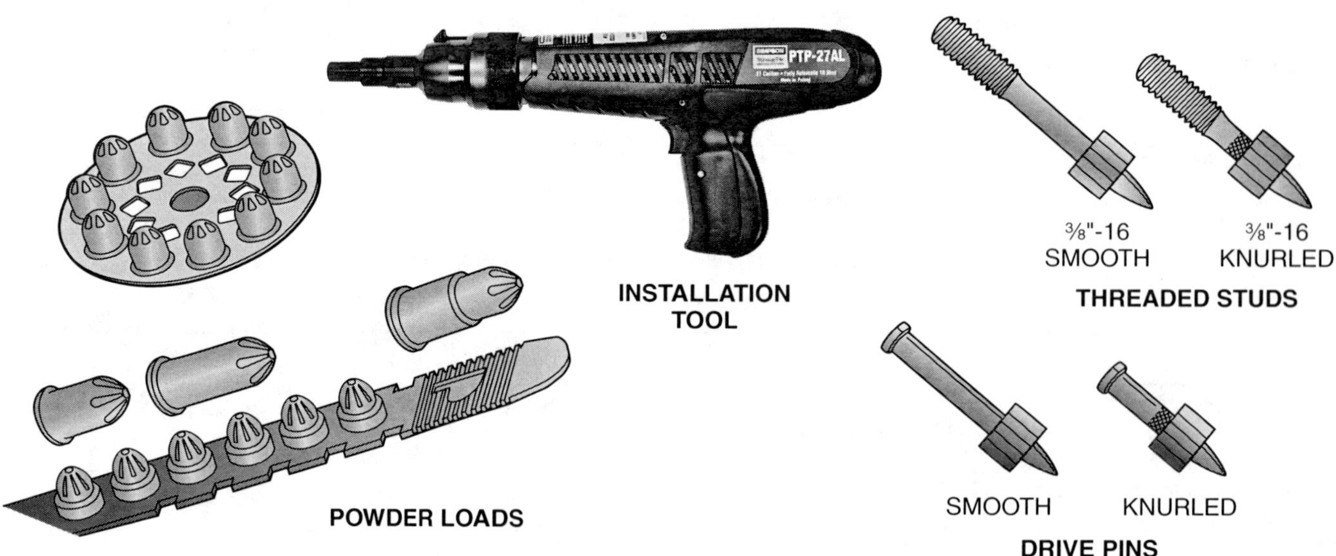

Figure 82 Powder-actuated installation tool and fasteners.

Powder-Actuated Fasteners

Powder-actuated fasteners can be used directly against a surface or with special extension rods for overhead or remote fastening in tight locations.

- Do not guess before fastening into any base material; always perform a center punch test.
- Always make a test firing into a suitable base material with the lowest power level recommended for the tool being used. If this does not set the fastener, try the next higher power level. Continue this procedure until the proper fastener penetration is obtained.
- Always point the tool away from operators and bystanders.
- Never use the tool in an explosive or flammable area.
- Never leave a loaded tool unattended.
- Do not load the tool until you are prepared to complete the fastening. If you decide not to make a fastening after the tool has been loaded, always remove the powder load first, and then the fastener. Always unload the tool before cleaning, servicing, or changing parts; prior to work breaks; and when storing the tool.
- Always hold the tool at a right angle to the work surface, and use the spall (chip or fragment) guard or stop spall whenever possible.
- Always follow the spacing, edge distance, and base material thickness requirements.
- Never fire through an existing hole or into a weld area.
- In the event of a misfire, always hold the tool depressed against the work surface for at least 30 seconds. If the tool still does not fire, follow the published tool instructions. Never carelessly discard or throw unfired powder loads into a trash receptacle.
- Always store the powder loads and the unloaded tool under lock and key.

7.0.0 PROJECT SCHEDULES

A construction project requires considerable planning and scheduling because different trades, equipment, and materials are needed at different times during the process. Electrical and communications cabling is normally installed when the building has been dried-in. This means that the exterior siding and roofing have been applied so that the building remains dry, but the framing is exposed in the interior of the building.

Project planning and scheduling will be covered in more detail later in your training. For now, *Figures 83* and *84* will give you an overview of where each trade fits into the construction process for residential and commercial projects. These figures provide a convenient way to compare the two schedules and to note the similarities and differences between them.

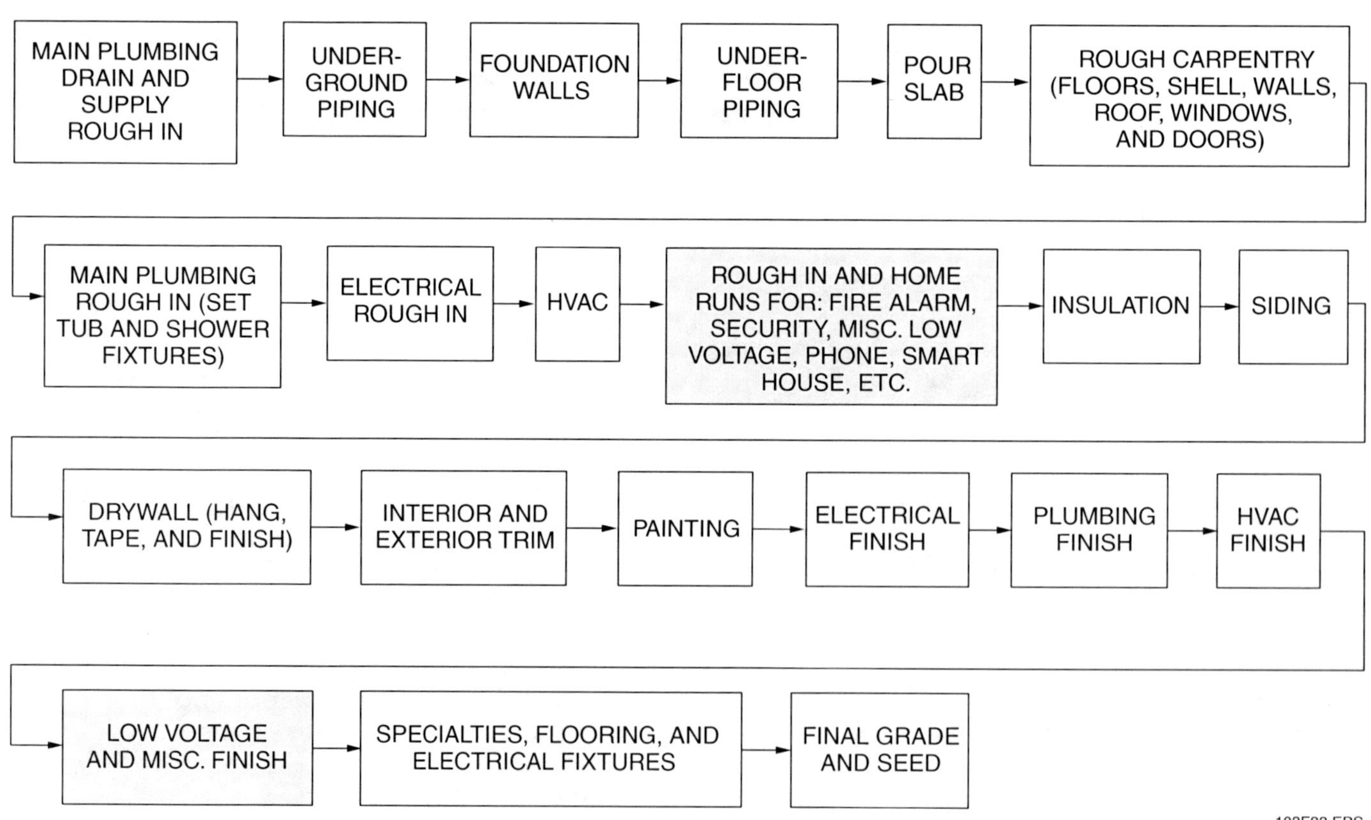

Figure 83 Typical residential construction schedule.

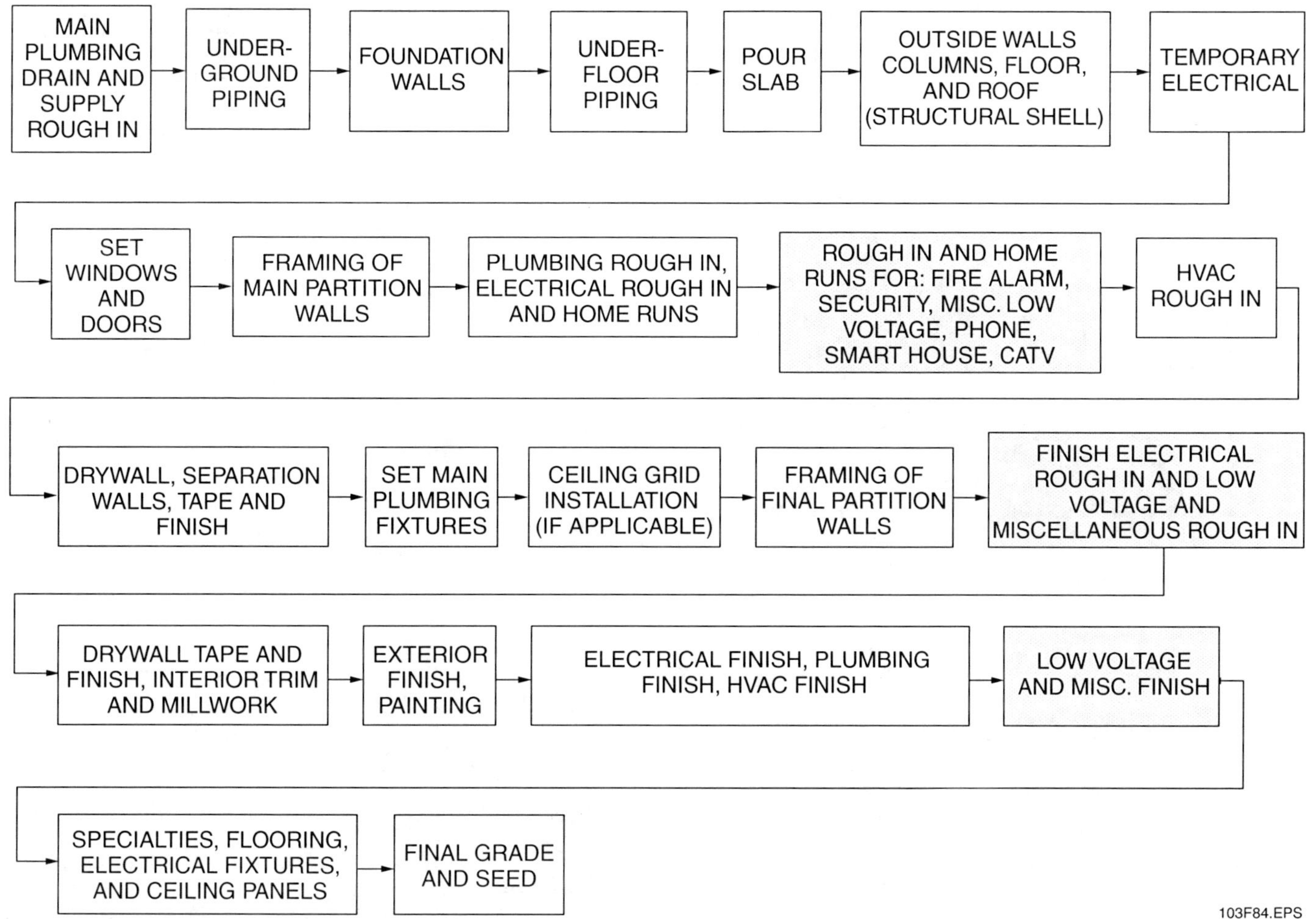

Figure 84   Typical commercial construction schedule.

SUMMARY

Concrete and steel are the main components of commercial structures, such as office buildings. The skeleton of the building can be made from either structural steel or concrete columns and beams. Floors are typically made of poured concrete over a corrugated sheet metal form. Cold-rolled steel studs are used to frame walls and partitions. Heavier-gauge steel studs are often used on loadbearing exterior walls. Prefabricated curtain walls of glass and metal are used to create the building façade.

The tools and fasteners used in wood-frame construction are generally not suitable for creating openings in concrete and steel structures. Special drills must be used to make the openings required to mount components and route cable through the building. An important skill to acquire in order to properly install cable systems is the ability to select the correct tool and fastener for the job.

1. Working with dry cement can be hazardous because it is highly flammable.

   a. True
   b. False

2. The standard steel stud used in nonbearing walls can range up to _____ inches wide.

   a. 1⅜
   b. 2½
   c. 4
   d. 6

3. Concrete tilt-up panels are typically _____ thick.

   a. 1" to 2"
   b. 2" to 3"
   c. 3" to 4"
   d. 5" to 8"

4. Drilling into post-tensioned concrete can cause problems because it _____.

   a. contains a steel cable
   b. is very brittle
   c. is extremely thick
   d. contains toxic material

5. In some applications, such as factories, the structure that may be found in the concrete floor to accommodate cabling and other services is a(n) _____.

   a. flushduct
   b. feeder duct
   c. channel
   d. trench

6. A modular floor raised on pedestals to allow cabling to be run underneath it is known as a(n) _____ floor.

   a. monolithic
   b. flushduct
   c. access
   d. elevated

7. The type of suspended ceiling panel that has a kerfed and rabbeted edge is used in a(n) _____ system.

   a. exposed grid
   b. concealed grid
   c. pan
   d. integrated

8. An integrated ceiling system normally uses _____ modules.

   a. 12" × 12"
   b. 12" × 24"
   c. 24" × 24"
   d. 30" × 60"

9. The wall shown in *Figure 1* is most likely located between _____.

   a. a doctor's office and a waiting room
   b. the garage and kitchen of a home
   c. a manufacturing area and an office
   d. a bedroom and a bathroom

10. An STC rating indicates _____.

    a. the degree of fire resistance of walls and ceilings
    b. how well a structure isolates airborne sound transmission
    c. the thickness of gypsum drywall
    d. how airtight the structure is

11. The quality of some fasteners can be determined by the _____ on the head.

    a. number of grooves
    b. number of sides
    c. length of the lines
    d. number of grade markings

12. The purpose of a jam nut is to _____.

    a. hold a piece of machinery stationary
    b. lock another nut in place
    c. stop the rotation of a machine for safety
    d. compress a lock washer into place

Figure 1

13. Wedge and sleeve anchors are both classified
as ______.

   a. bolt anchors
   b. hollow-wall anchors
   c. one-step anchors
   d. screw anchors

14. The type of bit that can cut through any type
of masonry is the ______.

   a. masonry twist bit
   b. percussion carbide bit
   c. type M flex bit
   d. hammer core bit

15. To drill a 6" opening in a reinforced concrete
wall, you would use a ______.

   a. hammering drill with a hole saw
   b. core drill with a core bit
   c. rotary drill with a chisel bit
   d. reciprocating saw with a concrete blade

Fill in the blank with the correct trade term that you learned from your study of this module.

1. The normal value used for stating the torque applied to a nut or bolt is ____________.

2. Concrete that is placed around cables that are subjected to tension once the concrete has hardened is called ____________.

3. The organization that publishes specifications and standards related to fasteners is the ____________.

4. The appropriate diameter of a bolt or screw is expressed as its ____________.

5. The turning force applied to a fastener is called ____________.

6. A mechanical fixture, such as a rigid tube that is used to protect the stripped end of a wire or fiber, is known as a(n) ____________.

7. A sealed chamber, such as the space above a suspended ceiling used as an air return, is called a(n) ____________.

8. Material added to concrete to obtain additional properties is referred to as a(n) ____________.

9. When parallel grooves appear in a surface design, it is said to be ____________.

10. The amount of deviation allowed from a standard is known as the ____________.

11. Another society that publishes specifications relating to fasteners is the ____________.

12. A groove or notch made by a saw is called a(n) ____________.

13. Concrete that has hardened but has not attained its full structural strength is called ____________.

14. The amount of space between the threads of a bolt and the threads of its nut is referred to as its ____________.

15. Concrete that is in a liquid or semi-liquid workable state is referred to as ____________.

16. Torque applied to small fasteners is often stated in ____________.

17. A board or panel that has a groove cut into one or more of its edges is said to be ____________.

18. A material that is formed with parallel ridges or grooves is said to be ____________.

Trade Terms

Admixture
American Society for Testing and Materials International (ASTM)
Clearance
Corrugated
Ferrule
Foot-pounds (ft-lbs)
Green concrete
Inch-pounds (in-lbs)
Kerf
Nominal size
Plastic concrete
Plenum
Post-tensioned concrete
Rabbeted
Society of Automotive Engineers (SAE)
Striated
Tolerance
Torque

Charles Wayne Adair

B & D Industries, Inc.
Project Manager and Communications Designer

As a young man, Charles Adair had the foresight to realize that he needed to channel the technical knowledge and experience he obtained while in the military into a field that had career potential in the commercial world.

What inspired you to enter the industry?

In 1970, while I was on active duty in the U.S. Air Force, I came to the realization that my career as a nuclear weapons specialist would not serve me well in the civilian arena, so I chose to cross-train into a different career ladder, one that was electronics based, and one that had a good future in the civilian economy. Electronic encryption equipment repair sure fit the bill. I originally got an electronics technician position working on electronic encryption equipment. From there, I moved into the communications arena. Communications is a field that has shown continuous and exciting leaps in technology over the last 40 years of my involvement. That technology is still evolving and will probably continue to evolve for the next 50 years or so.

What kinds of work have you done in your career?

Prior to high school graduation, I worked as a farmhand, a plumber's assistant, an electrician's assistant, carpenter's assistant, general construction laborer, and a shop mechanic's helper. In the military I worked as an electronics technician and nuclear weapons specialist. Outside the military I became an electronic encryption equipment technician, a communications technician, a video-teleconference field engineer, a communications circuit designer, and a project manager of construction work for communications cabling and equipment. Along the way, I have accumulated over 10 years of experience as a classroom instructor in the technical areas.

What do you enjoy most about your job?

As a communications designer, I enjoy seeing a concept become a reality. As an instructor, the pleasure of seeing an untrained individual become a competent and useful technician is one that I never fail to relish.

What types of training have you been through?

My first formal technical training was on the electronics used inside atomic weapons while in the U.S. Air Force. After I retired, I became a construction electrician. After about five years as a journeyman electrician, I again found employment in the communications field. About 10 years later, I started working with a company that specialized in video-teleconferencing and custom boardroom systems. I then got my BICSI technician-level certification and became a registered communications distribution designer (RCDD). I also received a certification as a certified fiber optic technician (CFOT).

How important is education and training in construction?

I have always considered proper education and training to be the key to success. I am convinced that my extensive training has made a big difference in where I am today. It is my firm belief that everyone must take the time to get all the formal and technical training they can, so they will not only be able to do the job, but will fully understand what they are doing and why it needs to be done a certain way.

What advice would you give to those new to the EST field?

Put yourself into your work. If you make no commitment, you will reap no rewards. Not all rewards are cash related; you can never put a price on job satisfaction.

Trade Terms Introduced in This Module

Admixture: Any material that is added to a concrete mixture to obtain additional properties.

American Society for Testing and Materials International (ASTM): An organization that publishes specifications and standards related to fasteners.

Clearance: The amount of space between the threads of bolts and their nuts.

Corrugated: Material formed with parallel ridges or grooves.

Ferrule: A mechanical fixture, usually a rigid tube, used to protect and align the stripped end of a wire or fiber.

Foot-pounds (ft-lbs): The normal method used for measuring the amount of torque being applied to bolts or nuts.

Green concrete: Concrete that has hardened, but has not yet gained its full structural strength.

Inch-pounds (in-lbs): A method of measuring the amount of torque applied to small bolts or nuts that require measurement in smaller increments than foot-pounds.

Kerf: A groove or notch made by a saw.

Nominal size: A means of expressing the size of a bolt or screw. It is the approximate diameter of a bolt or screw.

Plastic concrete: Concrete in a liquid or semi-liquid workable state.

Plenum: A sealed chamber for moving air under slight pressure at the inlet or outlet of an air conditioning system. In some commercial buildings, the space above a suspended ceiling often acts as a return air plenum.

Post-tensioned concrete: Concrete placed around steel reinforcements, such as rods or cables, that are isolated from the concrete. After the concrete has cured, tension is applied to the rods or cables to provide greater structural strength.

Rabbeted: The term applied to a board or panel that has a groove cut into one or more of its edges.

Society of Automotive Engineers (SAE): An organization that publishes specifications and standards related to fasteners.

Striated: A surface design that has the appearance of fine parallel grooves.

Tolerance: The amount of deviation allowed from a standard.

Torque: The turning force applied to a fastener.

Additional Resources

This module is intended to present thorough resources for task training. The following reference works are suggested for further study. These are optional materials for continued education rather than for task training.

Basic Construction Materials, 6th Edition. 2002. Theodore W. Marotta. Upper Saddle River, NJ: Prentice Hall.

Principles and Practices of Light Construction, 6th Edition. 2004. Ronald C. Smith, Ted L. Honkala, and Malcolm W. Sharp. Upper Saddle River, NJ: Prentice Hall.

Principles and Practices of Commercial Construction, 7th Edition. 2004. Cameron K. Andres and Ronald C. Smith. Upper Saddle River, NJ: Prentice Hall.

NCCER makes every effort to keep these textbooks up-to-date and free of technical errors. We appreciate your help in this process. If you have an idea for improving this textbook, or if you find an error, a typographical mistake, or an inaccuracy in NCCER's Contren® textbooks, please write us, using this form or a photocopy. Be sure to include the exact module number, page number, a detailed description, and the correction, if applicable. Your input will be brought to the attention of the Technical Review Committee. Thank you for your assistance.

Instructors – If you found that additional materials were necessary in order to teach this module effectively, please let us know so that we may include them in the Equipment/Materials list in the Annotated Instructor's Guide.

Write:	Product Development and Revision
	National Center for Construction Education and Research
	3600 NW 43rd St., Bldg. G, Gainesville, FL 32606
Fax:	352-334-0932
E-mail:	curriculum@nccer.org

Craft ___________________________ Module Name ___________________________

Copyright Date __________ Module Number __________ Page Number(s) __________

Description ____

____

____

____

(Optional) Correction ____

____

____

(Optional) Your Name and Address ____

____

____